PROTOPLASMATOLOGIA
HANDBUCH DER PROTOPLASMAFORSCHUNG

BEGRÜNDET VON

L. V. HEILBRUNN · F. WEBER
PHILADELPHIA   GRAZ

HERAUSGEGEBEN VON

M. ALFERT · H. BAUER · C. V. HARDING · W. SANDRITTER · P. SITTE
BERKELEY   TÜBINGEN   ROCHESTER   FREIBURG I. BR.   FREIBURG I. BR.

MITHERAUSGEBER

J. BRACHET-BRUXELLES · H. G. CALLAN-ST. ANDREWS · R. COLLANDER-HELSINKI
K. DAN-TOKYO · E. FAURÉ-FREMIET-PARIS · A. FREY-WYSSLING-ZÜRICH
L. GEITLER-WIEN · K. HÖFLER-WIEN · M. H. JACOBS-PHILADELPHIA
N. KAMIYA-OSAKA · W. MENKE-KÖLN · A. MONROY-PALERMO
A. PISCHINGER-WIEN · J. RUNNSTRÖM-STOCKHOLM

BAND V

KARYOPLASMA (NUCLEUS)

4

THE NUCLEAR STRUCTURES OF PROTOCARYOTIC ORGANISMS
(BACTERIA AND CYANOPHYCEAE)

1969

SPRINGER-VERLAG
WIEN · NEW YORK

THE NUCLEAR STRUCTURES
OF PROTOCARYOTIC ORGANISMS
(BACTERIA AND CYANOPHYCEAE)

BY

G. WOLFGANG FUHS
ALBANY, NEW YORK

WITH 86 FIGURES

1969

SPRINGER-VERLAG
WIEN · NEW YORK

ISBN-13: 978-3-7091-5589-9   e-ISBN-13: 978-3-7091-5587-5
DOI: 10.1007/978-3-7091-5587-5

TITEL-NR. 8733

Protoplasmatologia
  V. Karyoplasma (Nucleus)
   4. The Nuclear Structures of Protocaryotic Organisms
   (Bacteria and Cyanophyceae)

The Nuclear Structures of Protocaryotic Organisms (Bacteria and Cyanophyceae)

By

G. WOLFGANG FUHS

Albany, New York

With 86 Figures

---

[1] Review of published material for this article was completed on May 31, 1967. Results and illustrations for which no other source is given were obtained under contract with Deutsche Forschungsgemeinschaft, Habilitation Grant No. Fu 49/1.

Introduction

Bacteria and Cyanophyceae resemble other organisms in that they contain deoxyribonucleic acid (DNA) as their carrier of genetic information. The DNA complement in each cell effectively fulfills the functions of a nucleus and takes part in processes of genetic exchange involving recombination. The mechanisms of gene action resemble those in other organisms. Nuclear organization, however, is sufficiently different from the type commonly encountered in other members of the plant kingdom that the separation as a group of bacteria and Cyanophyceae is warranted. The recommended term P r o t o c a r y o t e s or p r o t o c a r y o t i c o r g a n - i s m s (preferred over P r o c a r y o t e s and p r o c a r y o t i c) reflects the more primitive status of nuclear organization in these forms as compared with all others (referred to as E u c a r y o t e s or e u c a r y o t i c organisms, CHATTON 1937, DOUGHERTY 1957, RIS 1961, STANIER 1961, FUHS 1965 d).

1*

The concept of a close relationship between bacteria and Cyanophyceae is not based on nuclear organization alone. Other aspects are membrane chemistry and the localization of photosynthetic and respiratory pigments in the cytoplasmic membrane and simple derivatives thereof. For details, the reader is referred to reviews by PRINGSHEIM (1949). STANIER (1961), STANIER and VAN NIEL (1962), and ECHLIN and MORRIS (1965). Equally well documented by these authors is the indication of close relationship which consists in the difficulty in defining a separating line between the bacteria and the Cyanophyceae. According to PRINGSHEIM (1953, 1960) photosynthetic bacteria and bacteria-like Cyanophyceae can be distinguished by the species of chlorophyll they contain. Among the non-photosynthetic forms, *Beggiatoa* resembles *Oscillatoria* sufficiently to be considered an apochlorotic blue-green alga (PRINGSHEIM); other forms do not offer such distinctive characters and cannot be classified conclusively. The question whether any fundamental difference exists in the nuclear organization of both types of organisms was raised by E. G. PRINGSHEIM some fifteen years ago. Currently the answer is negative, since recently acquired knowledge emphasizes the similarity (probably identity) of the basic organizational status of the nuclear material.

In the nuclear structures of protocaryotic organisms, the only structure-forming macromolecular species is DNA. Certain macromolecular but amicroscopical functional moieties may be permanently or temporarily attached to certain sites along the double-stranded molecule. The remainder of chemical species associated with the DNA consists of small molecules and ions. This statement also implies the absence of auxiliary structures such as a nuclear membrane and a mitotic apparatus; the absence of the latter renders protocaryotic organisms insensitive against colchicine (EIGSTI 1946, JAKOB 1950, PARVIS 1954, DE LAMATER et al. 1953, SCHWEISFURTH 1959, KUMAR 1965). The lack of a mitotic apparatus does not necessarily impair essential functions, since in protocaryotic organisms a complete genome may appear as a single linkage group, the physical basis of which is a coherent DNA molecule, the g e n o p h o r (CAIRNS 1963b, RIS and CHANDLER 1963). There has been some speculation that in some bacteria the genome may consist of several subunits, but the arguments produced are inconclusive as yet. In other bacteria, accessory genetic elements which are not attached to the bacterial genophor and which represent individual units of replication, apparently can be replicated in synchrony with the bacterial genome and can be transmitted to progeny cells in an orderly fashion. Such phenomena would call for a mechanism that like mitosis ensures the orderly distribution of sets of genetic elements into daughter cells.

Many bacteria and probably most small-celled Cyanophyceae under normal circumstances contain one genetic complement per cell, i.e. they are m o n o e n e r g i d i c. Some bacteria and probably most of the large-celled Cyanophyceae are p o l y e n e r g i d i c. One e n e r g i d, in this usage, represents one genetic complement plus a minimum quantity of indispensable cell constituents, membrane templates, etc., as would be necessary to form a viable cell (HARTMANN 1911, 1953, GRELL 1964). The number of energids per cell is assumed to equal the number of genetic complements as well as the number of viable cells that eventually may emerge from it without replication of its genetic material. (In eucaryotic cells, polyenergidy occurs in the form of cells containing several nuclei each or as polyploidy, referring to the co-existence of genomes within the boundaries of a single nucleus. Obviously terms such as "polynucleated" and "polyploid" are inappropriate for protocaryotic cells.) The number of energids per cell can be subject to variation as a response to certain environmental conditions or during certain phases of a developmental cycle.

The absence in protocaryotic nuclear bodies of structural components other than DNA markedly affects their structure and morphology. Since the protocaryon essentially is an accumulation of DNA, the amount, molecular organization and chemical state of the DNA are basic determinants of nuclear shape and fine structure. Therefore, the organized DNA molecule (the genophor) must be considered the principal subject of any treatise dealing with nuclear cytology in bacteria and Cyanophyceae.

The genophor presents itself as a DNA duplex approximately $2\,m\mu$ wide, i.e. $1/100$ to $1/1000$ as wide as the bacterial or cyanophycean cell. Its length, however, exceeds that of the cell by a factor of 100 or 1000, i.e. it amounts to 1 mm and over. Obviously, the arrangement of such a structure within the cell and its mode of replication are the major questions to be dealt with in the following sections. In the present context it is sufficient to note that the genophor hardly ever occurs in its maximum degree of dispersion, i.e. as a single fibre buried in the cytoplasm. The DNA is concentrated instead in certain areas of the cell. The shape of such areas, however, is highly variable and covers the range from a delicate network of branches to a compact nuclear body. Such regions, if seen in ultrathin sections in the electron microscope, are characterized by the presence of DNA as the only structural component and commonly are referred to as DNA - p l a s m (KELLENBERGER 1962: the term DNA-plasm is not to be confused with "nucleoplasm", the latter term designating the nuclear "sap" in eucaryotic cells). While a single DNA fibre is barely visible even in the electron microscope, the accumulated DNA of a single genophor forms a structure which is within the resolving power of the light microscope.

Since the genophor is a molecule, i.e. a coherent structure, the genetic material must correspondingly be organized in the form of coherent portions of DNA-plasm. Generalizing from currently available evidence we may also assume that any coherent portion of DNA-plasm represents at least one genome, i.e. one complete set of genetical information. Such nuclear entity representing a linkage group corresponds to a chromosome in eucaryotic cells and in all likelihood is homologous with the latter. Its fine structure and chemical makeup are more primitive than found in chromosomes, but its molecular organization renders it self-sufficient as to permit its operation as a (haploid) nucleus. These entities are referred to as n u c l e o i d s (PIEKARSKI 1937, 1950); in Cyanophyceae they are also known as c h r o - m a t i n e l e m e n t s (FUHS 1958).

The term "nucleoid" is a Greek-Latin hybrid and therefore is undesirable from a philologist's point of view. Unfortunately, a suitable substitute is not easily found. The logical alternative, the term "karyoid", has been used for entirely non-nuclear elements in algae and also for the poly-phosphate granules in bacteria and Cyanophyceae which in this context were confused with nuclear elements (Bringmann). This situation leaves the scientific writer with terms that are either inconvenient ("nuclear equivalent", "pro[to]caryotic nucleus") or inappropriate from a compara-tive cytologist's point of view ("chromosome", "nucleus"). The wide-spread indiscriminate use of the latter terms undoubtedly has aided in the recognition of bacteria as objects for genetical and cytological studies, but accumulating evidence not only does not support the need for the continued use of such terms but also has confirmed the nucleoids as objects for study in their own right which in some respects presently are better understood than their counterparts in eucaryotic cells. — It should be pointed out in passing that terms such as "acaryotic" appear similarly inappropriate, since nuclear structures and functions (which are undeniably present) are referred to in an exclusively negative sense.

While every nucleoid is a coherent structure, the opposite conclusion is not permissible, at least not on a cytological level. Since nucleoids are not bounded by a membrane, they may coalesce with adjacent nucleoids to form compound nuclear structures, the genetical valency of which cannot be determined by cytological methods except such that imply a quantita-tive determination of DNA on a cellular level. If the genetical valency of such structures is not known or is considered irrelevant, they should be referred to as n u c l e a r   b o d i e s ,   n u c l e a r   r e t i c u l u m or the like, depending on their appearance.

Several nucleoids operating in the same (polyenergidic) cell in some more or less perfect sort of coordination may be termed a   n u c l e a r   a p p a r a t u s   or   c h r o m a t i n   a p p a r a t u s , the latter expression being generally accepted for the compound nuclear structures of the higher Cyanophyceae (Spearing 1937). It appears that this type of coordinated nuclear behavior is expressed in only a few members of the bacteria and Cyanophyceae and that the more advanced types in the Cyanophyceae are distinctly different from those in bacteria. A more detailed account concerning this point will be made in the concluding section of this review after the facts have been presented in detail.

Nuclear Structures of Bacteria

I. Biochemical and Molecular Aspects

A. The Occurrence of Deoxyribonucleic Acid in Bacteria and Its Localization in Microscopically Visible Structures

Voit's (1925, 1927) macroscopic observation of a positive result with Feulgen's nuclear stain in thick smears and suspensions of various bacteria was the first conclusive demonstration of thymonucleic (deoxyribonucleic) acid in this group of organisms. Only much later and after many unsuccessful attempts, the same reaction was successfully applied on the microscopic level, and it revealed the regular occurrence of DNA-containing structures in bacteria (STILLE 1937, PIEKARSKI 1937, NEUMANN 1941). The localization of DNA in "nucleus-like" bodies was later confirmed with the aid of specific enzymes (TULASNE and VENDRELY 1947 a, b, TULASNE and MINCK 1947, BOIVIN et al. 1947 a, LEE 1960) and by autoradiography (CARO 1961, VAN TUBERGEN 1961).

The demonstration with biochemical methods of DNA in bacteria was first accomplished by the identification of its constituents, deoxyribose, thymine, adenine, guanine, and cytosine (JOHNSON and BROWN 1922, EPSTEIN et al. 1936, BELOZERSKIJ 1939). More recently, the isolation of DNA from bacteria in more or less depolymerized but otherwise intact form has become a routine procedure. The base compositions of the various samples obtained suggest that native bacterial DNA in accordance with the Watson and Crick model forms helical two-stranded molecules with bases arranged in guanine-cytosine and adenine-thymine pairs and some "minor" bases taking the places of their structural analogues among the former. Guanine-cytosine pairs account for 25 to 75 per cent of the base pairs in bacterial DNA (for compilations of data see BELOZERSKIJ and SPIRIN 1960, McCARTHY 1965 and HILL 1966).

B. Taxonomical Implications of DNA Biochemistry

Similarities of DNA base ratios in similar bacteria have led to attempts to use base ratios as taxonomic characters. Information obtained in this way, however, is only indicative, since the genetic message is determined by the base s e q u e n c e rather than the overall base ratio. Even the opposite conclusion, i.e. that almost identical messages in related strains are reflected in almost identical base ratios of the respective DNA's does not always hold, since considerable variation of base ratios is known to occur also among certain apparently closely related strains (see references in preceding section). The same holds for several types of artifically induced mutants characterized by the non-utilization of sugars or by respiratory deficiencies, which show base ratios quite different from those of the parent strains (SPIRIN et al. 1958a, WEED 1963, DE LEY 1964, GAUSE et al. 1964). Whether this phenomenon can be explained in terms of degeneracy of the genetic code is not clear at the present time.

Experiments on the reversible denaturation of DNA (DOTY et al. 1960, MARMUR and LANE 1960, MARMUR and DOTY 1961, MARMUR et al. 1961 b) have resulted in the development of a technique which permits the detection of homologous regions in any two species of DNA, provided the genetic complement is relatively small as in bacteria and viruses. The method consists in "annealing" a mixture of the DNA's in single-stranded form and quantitative determination of the helical material formed. The method represents a straight-forward approach to the comparison of genetic messages and already has provided a considerable amount of valuable information (MARMUR 1963, DE LEY 1964, DE LEY and PARK 1966, PARK and DE LEY 1967, ROGUL et al. 1965). DUBNAU et al. (1965) found marked base-sequence homologies in ribosomal and transfer ribonucleic acids of several *Bacillus* strains, probably indicating the existence of a common "genetic core" of closely related species.

C. The DNA Content of Bacterial Cells

The amount of DNA in normal cells of Eubacteria and forms of comparable size usually is in the range of 1.0×10^{-15} to 2.0×10^{-14} grams per cell. In young cells of *Escherichia coli* B and B/r GILLIES and ALPER (1960) found

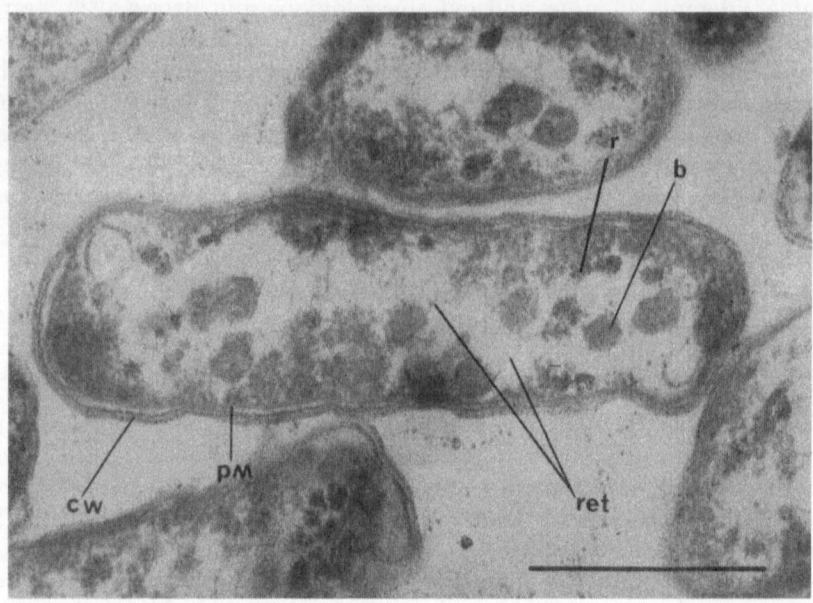

Fig. 1. Electron micrograph of sectioned *Ferrobacillus ferrooxidans*. The specimen was fixed and handled following the procedure of KELLENBERGER, RYTER and SÉCHAUD (1958 a). Epon 812 was the embedding polymer, and sections were cut with glass knives and examined in an RCA EMU-2D electron microscope. *CW*, layered cell wall; *pm*, cytoplasmic membrane; *b*, dense membraneous bodies believed to be mesosomes; *r*, ribosomal cluster; *ret*, irregular reticular network in the transparent region is the DNA-plasm of the cell. Scale marker: 0.5 µ. — Reprinted from DUGAN and LUNDGREN (1964).

1.37 and 1.82×10^{-14} grams respectively, and 0.78 and 1.28×10^{-14} grams respectively in cells from 24 hour cultures. Somewhat higher values were reported by SPIRIN et al. (1958b). HERSHEY and MELECHEN (1957) found 0.65×10^{-14} grams of DNA in *E. coli* cells growing at a generation time of 1 hour. *Bacillus subtilis* spores were found to contain 5×10^{-15} grams equalling 3.0×10^9 molecular weight units of DNA; the DNA content of vegetative cells was two or three times as high (DENNIS and WAKE 1966). In comparative studies, spores of other bacilli showed a DNA content two, three or four times the amount found in the above species (FITZ-JAMES and YOUNG 1959). EBERLE and LARK (1967) arrive at a slightly higher estimate of the non-replicating *B. subtilis* genome (3.32 and 3.9×10^9 molecular weight units).

DNA usually accounts for 1 to 10 per cent of the cell dry weight. In the course of early investigations on the relationship between nucleic acid content and protein synthesis of bacteria, CALDWELL and HINSHELWOOD (1950) found little change in the DNA content of *Aerobacter aerogenes* under a variety of environmental conditions, if DNA content was expressed on a cell nitrogen basis. Contrary to DNA, the amount of ribonucleic acid (RNA) varied with growth rate (CALDWELL et al. 1950).

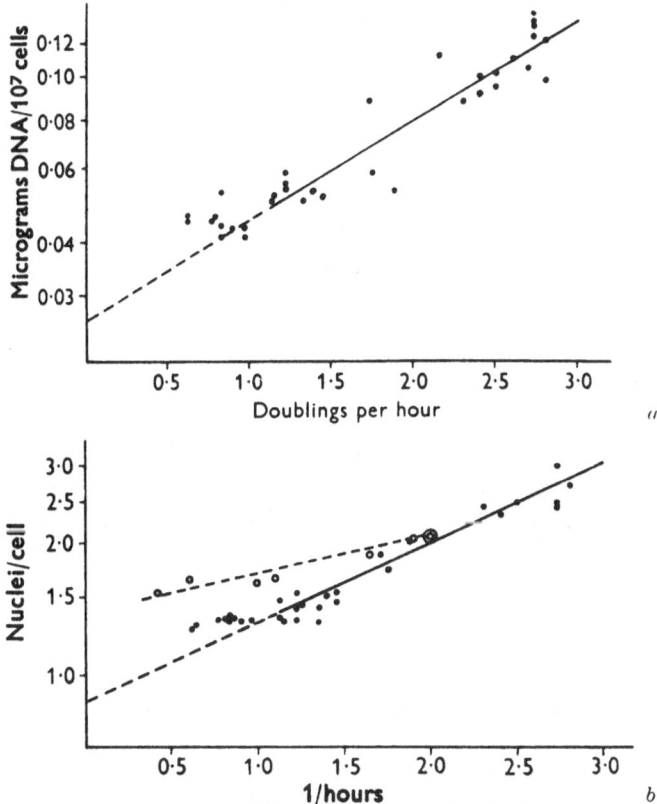

Fig. 2 *a*. Dependency of cellular deoxyribonucleic acid on growth rate at 37° C. The DNA content of the cultures was calculated from the deoxyribose determinations (μg DNA = μg deoxyribose × 2.44). The logarithm of the DNA values in μg/10⁷ viable cells is plotted against the growth rate. — Reprinted from SCHAECHTER et al. (1958).

Fig. 2 *b*. Dependency of the number of nuclear bodies per cell on growth rate at 37° C. The average number per cell was calculated from direct counts on stained preparations. The logarithm of the values is plotted against the corresponding growth rate (●). The stippled line corresponding to the open circles and the double-ringed point represent results from continuous culture experiments, plotted against dilution rate (○). — Reprinted from SCHAECHTER et al. (1958).

As to RNA, these findings concur with others according to which RNA content parallels rate of protein synthesis and rate of growth on a mass or per cell basis (MALMGREN and HEDÉN 1947, KJELDGAARD 1961, and others).

As to the DNA, the situation has proved to be more complex. During periods of extremely slow growth or starvation, cells may contain relatively little RNA and protein as compared with DNA, and the DNA-plasm may occupy the larger part of the cell (Fig. 1). On the other hand, during periods of accelerated growth, cell divisions tend to lag with respect to DNA synthesis, and the amount of DNA per cell increases.

At first this phenomenon was recognized as a change in the number of nuclear bodies per cell at certain points of the growth cycle (Badian 1933). Piekarski (1937) similarly found cells with two to four nuclear bodies per cell in actively growing cultures ("primary forms") and cells with one or two nucleoids only in older, stationary cultures ("secondary forms"). But it was not until much later that a positive straight-line correlation was established between growth rate and the number of nuclear bodies as well as growth rate and DNA content per cell, growth rate in this case being determined by the composition of the medium, not by temperature (Fig. 2) (Schaechter et al. 1958, see also Dean and Hinshelwood 1960 and Dean 1962). This suggests that the changes observed by Badian and Piekarski might be the result of the change in growth rate during normal development of an agar or liquid batch culture.

There are, however, phenomena basically resembling these but difficult to explain in purely quantitative terms. One of these is the formation of very long multinucleated swarmer cells in *Proteus spp.* which originate from small cells containing one nuclear element each. Later the swarmers divide into small uninucleated cells which upon transfer again may give rise to swarmers (Fig. 3, Hoeniger 1966). Complicated processes such as formation of spores or other resting stages involve the reduction of the nuclear material to a minimum (for details see Section II, G. 3, p. 73).

A more elaborate type of nuclear organization also appears to exist in the large forms described by Delaporte (1964a, b) which are polyenergidic, since their endospores receive only a minor portion of the nuclear material. The number of genetic complements in the vegetative cells apparently is not determined by growth rate alone but subject to more rigid control. Controlled polyenergidy of another type is found in the larger Cyanophyceae (p. 136).

The minimum amount of genetic material obviously is represented by a complete set of genetical information (one g e n e t i c   c o m p l e m e n t or n u c l e o i d, see introduction). As pointed out in considerable detail below, the amount of DNA equivalent to one such genetic complement was found, in the case of *E. coli* strain K 12, to equal 4.2×10^{-15} grams or 2.5×10^9 molecular weight units (Cairns 1963b). Cooper and Helmstetter (1967) arrive at an identical estimate for the genome of *E. coli* strain B/r. A higher DNA content per cell either is due to the fact that the DNA is in a state of replication, or that several complements are present, or both.

This approach permits the conversion of DNA mass units into numbers of genetic complements per cell. Accurate determinations of DNA content and its rate of synthesis are an essential part of this approach and have been greatly facilitated by the advanced techniques of ^{3}H-thymidine labeling and related techniques (K. G. Lark and C. Lark 1965 and related papers). C. Lark (1966) found variation of growth rate in *E. coli* 15 T$^-$ between 40 and 270 minutes on various carbon sources to be correlated with DNA contents per cell between 1.38 and 0.51×10^{-14} grams. This corresponds with an average content of 1.2 to 3 genetic complements per cell. Eberle and Lark (1967) present a corresponding set of data for *B. subtilis* indicating that this form contains between one and two replicating genomes per cell.

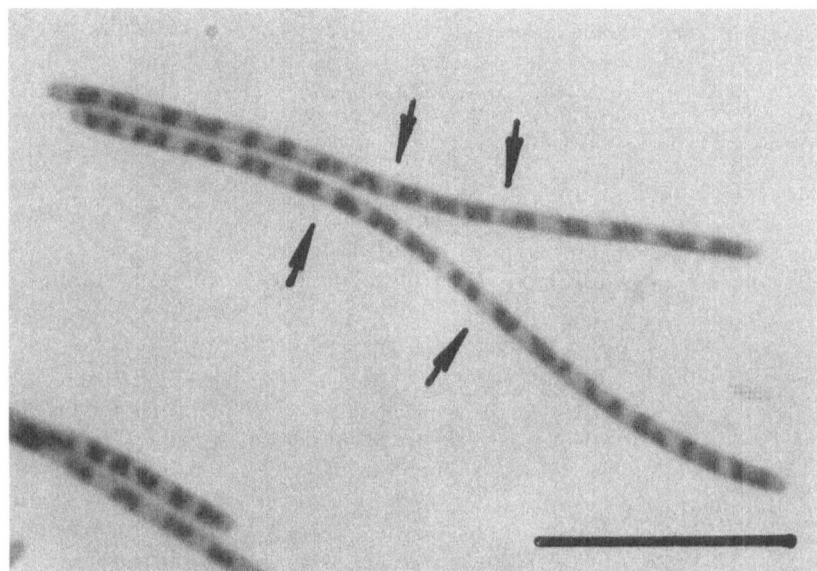

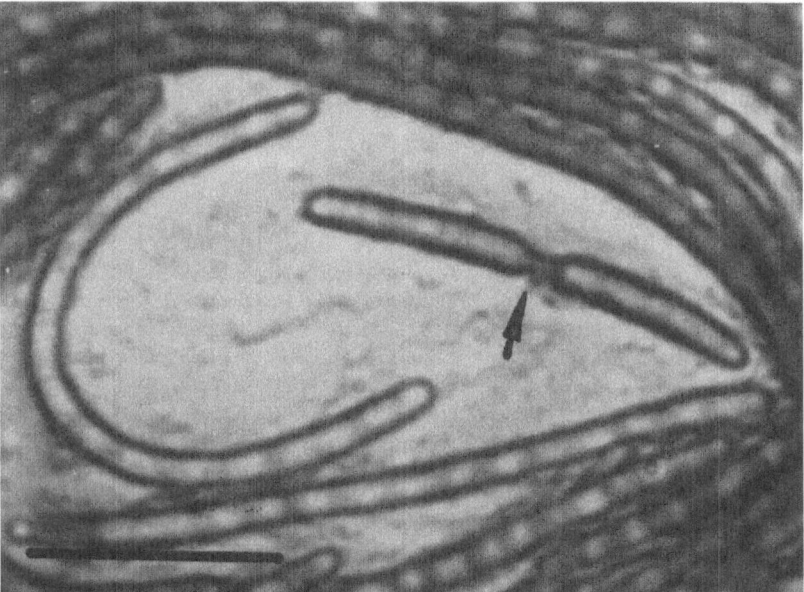

Fig. 3 *a* and 3 *b*. Staining of nuclear bodies and cell walls respectively in swarmers of *Proteus mirabilis*. Arrows in Fig. 3 *a* point to stained bands between nuclear bodies; arrow in Fig. 3 *b* indicates point at which swarmer is dividing. Nuclear stain: osmium-HCl Giemsa, counterstained with 0.01 percent basic fuchsin, mounted in dilute Giemsa. Cell wall stain according to HALE (1953), modified. Scale markers indicate 10 μ. — Reprinted from HOENIGER (1966) with permission, Canadian Journals of Research.

D. Molecular Size and Circularity of Bacterial DNA

Numerous attempts have been made to isolate bacterial DNA with conventional biochemical techniques and to determine its molecular weight. During these studies it became evident that chain length varied with the procedure used and that the apparent molecular weight of native DNA in solution could be significantly decreased by mechanical shearing, and that shearing during common laboratory procedures such as stirring, pipetting, capillary viscosimetry and forced sedimentation was sufficient to cause such damage. A molecular weight of 250×10^6 was the highest found with bulk isolation techniques (MASSIE and ZIMM 1965).

Two elegant methods have been developed that permit isolation of essentially unbroken molecules of bacterial DNA. One of these (KLEINSCHMIDT et al. 1960a, b, 1961) consists in lysis of protoplasts at an air-water interface and simultaneous incorporation of the liberated DNA into a monomolecular film of a basic protein, cytochrome c. The film is produced in a Langmuir trough, and after slight compression small areas of the layer are picked up with an electron-microscopic grid. The specimen is shadowed with platinum-carbon on a rotating stage under a small angle. The electron micrographs (Fig. 4) show the DNA from a single protoplast spread into a two-dimensional pattern. The most conspicuous feature in such preparations is the apparent absence of free ends, kinks, and ramification of the molecular fibre. It appears that the entire DNA pool of the protoplast is made from a single unbranched DNA molecule. Quite commonly the pool is spread into loops originating from a central piece of debris which is considered a remnant of the protoplast.

A different technique was developed by CAIRNS (1962a, b, 1963a, b). The DNA of thymine-requiring strains of E. coli was labeled with ³H-thymidine. The bacteria were then lysed in the funnel of a membrane filter apparatus together with an excess of identical but non-labeled material which served to "dilute" the labeled material and to reduce shearing forces during subsequent manipulation which merely consisted in the gentle deposition of the suspended material onto the filter surface. After drying, the filter was coated with a nuclear emulsion, exposed and developed. The autoradiograms were investigated with the aid of a light microscope (Fig. 5). In this way, silver grain contours of the DNA fibres produced by tritium decay are observed rather than the DNA molecules themselves. The majority of the DNA pools was in the form of heavily entangled aggregates as shown in Fig. 4. In the autoradiograms, such formations appear as small aggregates of silver grains which do not reveal significant details. A few per cent of the DNA pools, however, were completely unfolded and accessible to analysis. (Contrary to the light microscope, the electron microscope would not permit a complete survey of the unfolded structure because of the limitation in size of unsupported specimen area.)

As a result of these investigations it was found that DNA molecules in E. coli were as long as 1100 to 1400 μ (CAIRNS 1963b, BOURGUIGNON 1964). The two-strandedness of this material was evident from the number of silver

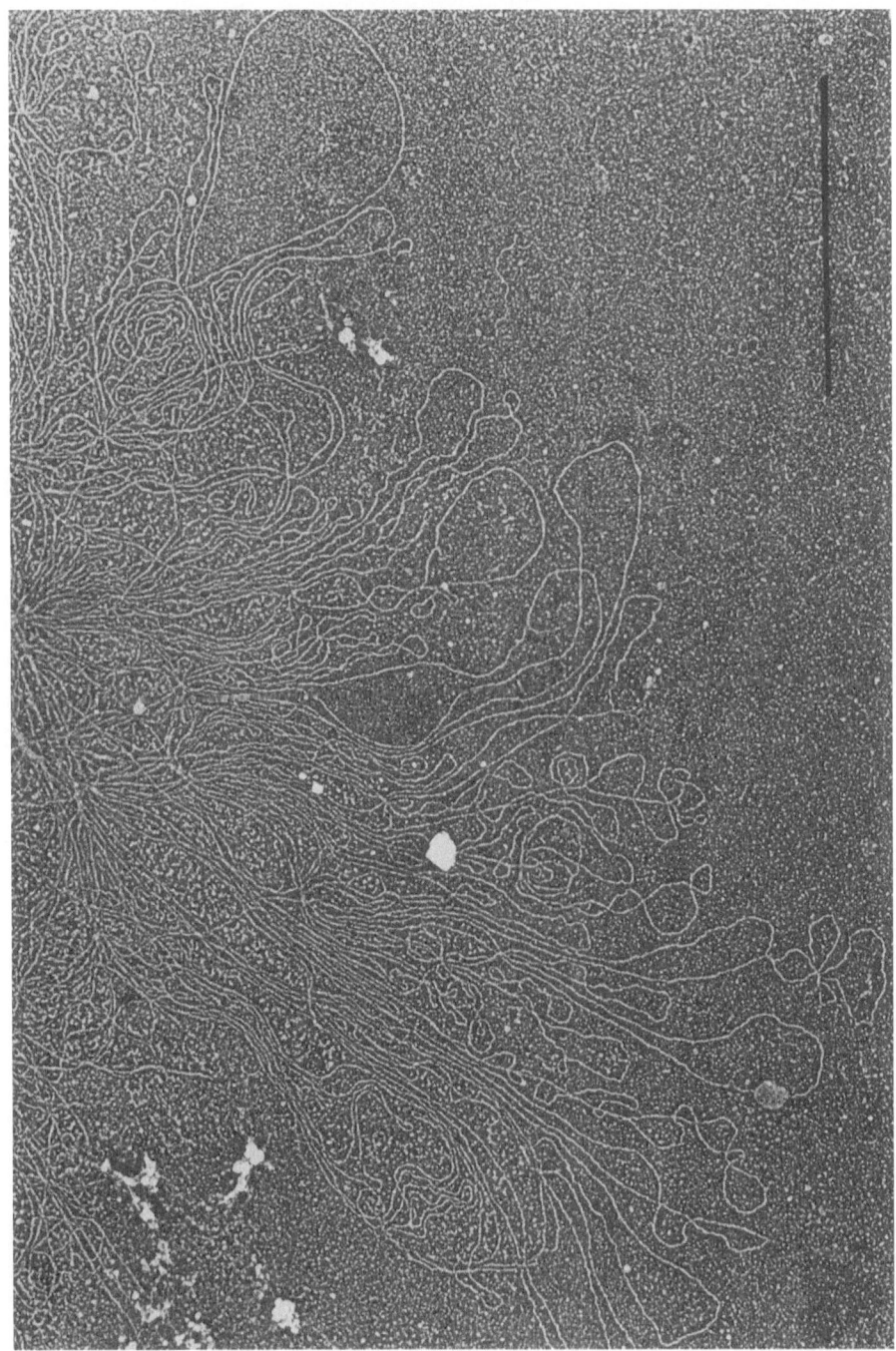

Fig. 4. Lysed protoplast of *Micrococcus lysodeikticus*, spread on air-water interface in a cytochrome-c monolayer. Rotating specimen shadowed with platinum-carbon under 6 to 10 degree angle. Scale marker indicates 1 μ. — Reprinted from KLEINSCHMIDT et al. (1961).

grains per unit length of fully labeled material which was identical with
that obtained with bacteriophage lambda, the DNA of which was known to
be two-stranded (Cairns 1962a, b).

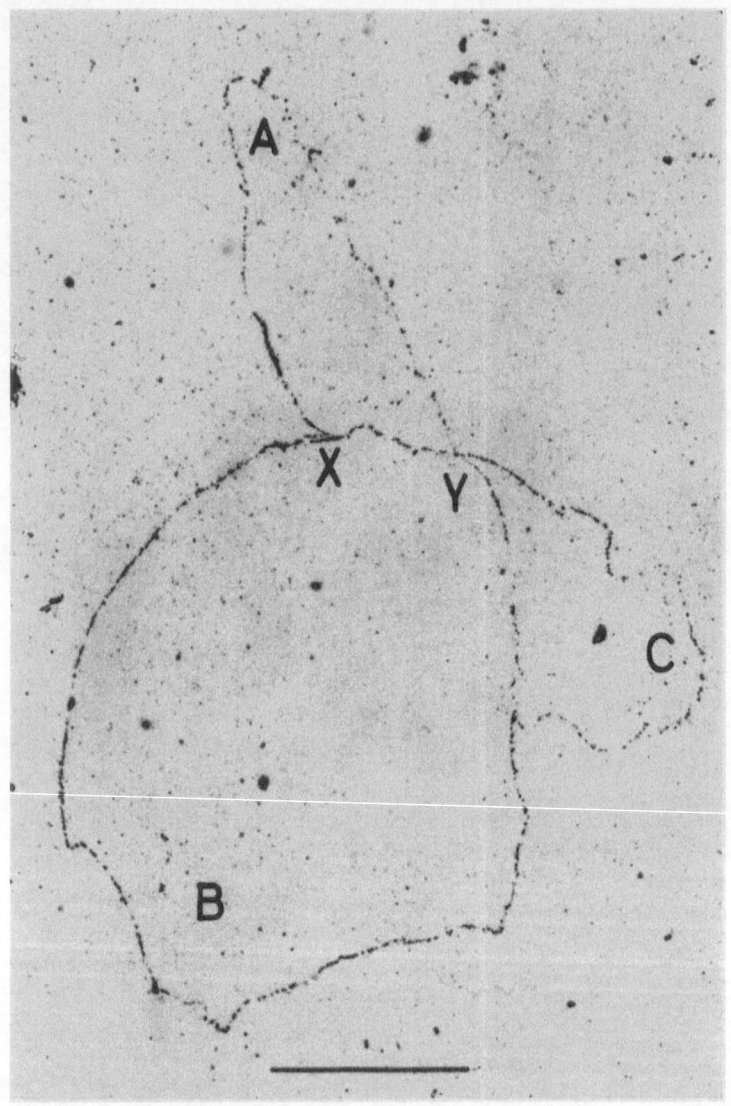

Fig. 5. Autoradiograph of the genophor of *E. coli* K 12 Hfr, labeled with tritiated thymidine for two generations
and extracted with lysozyme. The genophor has completed approximately two thirds of its replication cycle. The
starting and finishing point of replication is marked *X*, the replication fork *Y*. *A* and *B* are the replicated parts,
part *C* is still unduplicated. Since rounds of duplication and the period of labeling do not coincide, the unreplicated
part is partly half-hot (*Y—C*) and partly hot-hot (*C—X*). The scale marker represents 100 μ. — Reprinted from
Cairns (1963 b).

It was also observed (Cairns, loc. cit.) that *E. coli* DNA could exist as a
circular structure. This was an important verification of predictions based
on genetic experiments (Wollman and Jacob 1959).

The amount of DNA contained in these circular structures was of the same magnitude as the minimum DNA complement of a viable *E. coli* cell. It cannot be doubted therefore that these structures represent the complete genome of *E. coli* K 12 which was known to form a single circular linkage group. Thus the observed structures represent the organized DNA molecule which is the carrier of genetic information and which was referred to in the introduction as the g e n o p h o r.

It is not clear at the present time whether bacterial genomes in general are organized as single coherent DNA molecules. For *B. subtilis* the existence of several molecular subunits has been suggested but evidence is still incomplete and conflicting (GANESAN 1963, GANESAN and LEDERBERG 1965a, KELLY and PRITCHARD 1965, DENNIS and WAKE 1966).

E. Replication of Bacterial DNA

1. The Semiconservative Mode of Replication

In their comments on the basic architecture of the two-stranded helical DNA molecule WATSON and CRICK (1953 a, b) speculated that during replication the helix might be uncoiled and that the polynucleotide strands might become separated, each serving as a template for one of the two new complementary strands. This mode of replication in which the daughter molecule consists of one parental and one newly synthesized strand was termed "s e m i c o n s e r v a t i v e" by DELBRÜCK and STENT (1957).

Shortly thereafter evidence began to accumulate which proved that this model applied to the mode of replication of bacterial DNA. MESELSON and STAHL (1958a, b) subjected crude preparations of *E. coli* DNA to equilibrium centrifugation in a caesium chloride gradient. The bacteria had been grown on $^{15}NH_4Cl$ as sole source of nitrogen long enough to contain almost exclusively ^{15}N-labeled DNA and subsequently had been transferred to $^{14}NH_4Cl$-containing medium for exactly one, two or three generations. After one generation all DNA was found to be of intermediate density as one would expect of DNA containing 50 per cent of each ^{15}N and ^{14}N atoms ("hybrid DNA"). After two and more generations, ^{14}N-DNA ("light DNA") appeared in increasing amounts besides a constant amount of intermediate-density DNA. The results are in accordance with the following model (Fig. 6).

Upon heating to 100° C, which is known to cause strand separation in native DNA, the hybrid material separated into heavy and light fractions of equal size which showed all properties of denatured (single-stranded) DNA (MESELSON and STAHL 1958a, b, DOTY et al. 1960, BALDWIN and SHOOTER 1963, FREIFELDER and DAVISON 1962). From these experiments it was concluded that the conserved units (shown as bars in Fig. 6) represent polynucleotide strands and that the experiment described in the preceding paragraph was valid proof of the semiconservative mode of replication of bacterial DNA.

Since the preparations used in these experiments consisted of fragmented, low molecular weight DNA, no conclusion could be reached whether the observed pattern of segregation and strand preservation was also valid for the entire bacterial genophor. An affirmative answer concerning this point was obtained by ultrasonic treatment of hybrid DNA (Rolfe 1962) and in more definitive form by autoradiography of microcolonies grown from single cells labeled with ^{3}H-thymidine (Painter et al. 1958, Forro and Wertheimer 1960, van Tubergen and Setlow 1961, Herman and Forro 1964, Forro 1965). As a rule, the first cell division produced equal division between daughters of the parental label. The segregation pattern in the second generation varied. The third division resulted in an all-or-none distribution of label, and this pattern governed all subsequent generations. From these observations it was concluded that *E. coli* cells contained between two and four large conserved units of DNA. Actually, in a random population, the cells contain also unfinished replicates of such units, and proper allowances have to be made for these (Herman and Forro 1964). Since the conservation of genetic material at replication takes place on the level of the deoxyribonucleotide strands, one may assume that the large conserved units of Forro et al. are structures on that level of organization, i.e. half-molecules of DNA. This conclusion is also supported by observations on the number of silver grains produced in autoradiograms by a known quantity of fully labeled DNA.

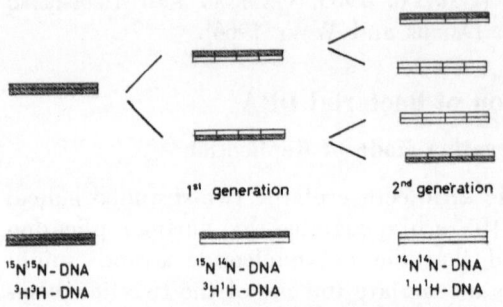

1st generation   2nd generation

^{15}N^{15}N - DNA ^{15}N ^{14}N - DNA ^{14}N^{14}N - DNA
^{3}H^{3}H - DNA ^{3}H^{1}H -DNA ^{1}H^{1}H - DNA

Fig. 6. The semiconservative replication of bacterial DNA and the bacterial genophor. Diagrammatic representation of two-stranded DNA molecules and of the distribution of label from a parent molecule into two generations of daughter molecules. Segmentation of the molecules indicates (a) the results obtained with fragmented bacterial DNA by Meselson and Stahl (1958 a, b), and (b) the validity of the model for entire genophors as demonstrated in other types of experiments (see text).

In reality, however, conservation of DNA units is not quite as rigid as might appear from the preceding paragraph, but on the other hand, randomization is not sufficiently frequent to support the idea of an alternate major mechanism of replication although being somewhat more frequent than anticipated on the basis of frequency of genetic recombination. Forro (1965) found 0.5 to 0.7 exchanges of strands per duplex per generation. Witkin (1951) and Messer (1963) presented results of quite different nature suggesting randomization of genetic material at a moderate rate.

The analysis of the pattern of phenotypic expression in *E. coli* mutants by Witkin (1951) produced results indicating the existence of two to four conserved units of DNA per cell or two conserved units per nuclear body (nucleoid).

Additional evidence concerning this point was independently obtained by experimental incorporation into bacterial DNA of radioactive phosphorus (^{32}P) and the subsequent analysis of the rate of killing in such population during storage at $-196°$ C (Fuerst and Stent 1956, see Stent and Fuerst 1960 for review). The shape of the survival curves suggested that

E. coli cells contained about three DNA units. The preservation of one of these units was required for survival of the cell. Experiments involving starvation of thymineless and normal cells showed that the DNA was the site of damage by ^{32}P-decay, and that the number of sensitive units per cell paralleled the number of nuclear bodies. Maximum sensitivity was reached by unlabeled cells after 1.6 generations in ^{32}P-containing medium, and fully labeled cells became less susceptible after one to two duplications in cold medium. This compares well with the appearance of "light" DNA in the experiments by MESELSON and STAHL and with the appearance of unlabeled progeny cells in the autoradiographs of microcolonies as described above.

CAIRNS (1962a, b, 1963a, b) presented results which lend strong support to the ideas expressed above. His findings are reported in the following section.

2. Sequential Replication of DNA

The autoradiographic analysis of isolated DNA pools was mentioned earlier (Section I, D, p. 12–14). By properly timing the addition and withdrawal of the labeled precursor, ^{3}H-thymidine, it was found that upon exposure for a fraction of one generation, one segment of newly labeled DNA per pool appeared, indicating that replication took place at a single point only. The newly labeled segment represented a fraction of the genophor which was identical with the fraction of the duplication time the cell had been exposed to the labeled precursor. This and the configuration found in the autoradiographs (Fig. 5) was proof that the replication of the bacterial genophor starts at one point only (the o r i g i n) and proceeds from there with uniform speed. Under the conditions chosen, replication occurred throughout the cell cycle with no apparent intermission. Furthermore, the *E. coli* genophor maintained its circular shape while being replicated. The essence of these findings by CAIRNS (1963a, b) is summarized in a schematic representation (Fig. 7).

It may be recalled that the autoradiographic technique did not permit an analysis of all DNA pools of a given population. Actually only very few of the pools existed in a completely unfolded form. Other techniques had to be used to establish the general validity of the proposed model. The existence of one replication point only per *E. coli* genophor is in agreement with results obtained with a combination of ^{14}C and density labeling techniques (BONHOEFFER and GIERER 1963).

Important confirmation and correlation with genetical data was obtained in the determination of frequency patterns of genetic markers and prophage attachment sites in growing bacterial populations. At any given time of the cell cycle, a marker or prophage located close to the origin has doubled more likely than another in a more distant location. YOSHIKAWA and SUEOKA (1963) used transformation for marker assay in *B. subtilis* and found different frequencies of certain markers in growing and resting populations. Assuming that resting cells were in a state of having completed a round of DNA replication, the relative frequency of any one marker in the growing population compared with its frequency in the stationary culture was taken as a measure of its proximity to the origin. The various markers then were arranged in the order of their relative frequency, and the sequence obtained was in good agreement with known frequencies of recombination. An analysis of marker frequencies in *B. subtilis* during the first few generations following spore germination gave compatible results (Wake 1963). In this period, DNA synthesis and cell divisions are known to be synchronous to a certain extent (Young and Fitz-James 1959). In another attempt, Jyssum (1965) found polarity of genophor replication in *Neisseria meningitidis*.

Nagata (1962, 1963 a, b) assayed the number of prophage sites in synchronously growing populations of *E. coli* K 12 Hfr and found their number to double at certain points of the cell cycle. Despite the use of a somewhat arbitrary reference point (cell division) several important details were borne out: First, different strains which at conjugation transfer genetic markers in a different sequence but with identical polarities (i.e. in sequences obtainable by cyclic permutation) replicate identical prophage sites during different periods of their cell cycles. A relation was found to exist between the time of replication and the time of transfer of such sites at conjugation. Second, the sequence of replication of two d i f f e r e n t prophage sites is opposite to the sequence of their transfer at conjugation. Obviously replication and transfer proceed with opposite polarities. Third, synchronous cultures of F⁻ strains of *E. coli* K 12 unlike Hfr and *B. subtilis* strains failed to replicate prophage sites in a synchronous fashion. The most likely explanation is that in cells of an F⁻ population the origin of replication of the DNA is located in different positions along the genetic map. If such population is raised from a single cell, randomization of the site of the origin must occur fast enough to become efficient within the period of growth into a manageable population (10 generations and over).

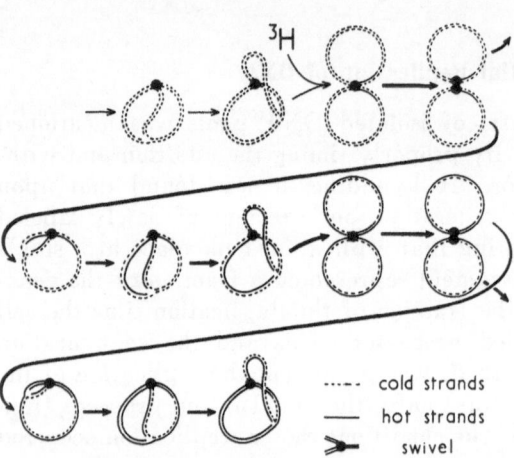

----- cold strands
——— hot strands
➤ swivel

Fig. 7. Diagrammatic representation of the replication of a circular bacterial genophor (*E. coli* K 12 Hfr thy⁻). Each genophor is shown as a circular double line representing a two-stranded molecule. Replication proceeds in a semiconservative manner. DNA synthesis is indicated as growth of "hot" strands after addition of ³H-thymidine to the growth medium. The "swivel point" is intended to indicate the requirement for a mechanism that permits the unduplicated part to rotate around its axis (see Section I, E, 3, Uncoiling the Helix, for details concerning this point). — Reprinted from Cairns (1963 b).

Finally, Vielmetter and Messer (1964) used the mutagenic effect of ³²P together with its ability to produce double strand breaks in DNA. Such breaks are "lethal" for the genophor except when replication is in progress a n d already has proceeded beyond the point where the double-strand break occurs—in this case, one of the daughter duplex molecules will "survive", i.e. will enable the cell to remain viable. Also, occasional single-strand mutational events caused by ^{32}P decay would encounter a smaller chance to be replicated during any replication cycle when located at a greater distance from the origin. Phenotypic expression of the former would be delayed with respect to that of the latter by one generation, i.e. the former would mainly appear as heterozygotes forming sectorized colonies, while the latter would have a greater tendency to appear as pure clones. The ratios of heterozygotes and pure mutant clones obtained for various markers were found to be principally in agreement with the predictions and with NAGATA's interpretation of the replication patterns in Hfr and F− strains of E. coli K 12.

3. Unwinding the Helix

The alleged continuity of the bacterial genophor as a helical two-stranded molecule raises the question as to the unwinding of its more than 400,000 turns within the short period of a single replication cycle. If the parental portion of the molecule would remain stationary, the "daughters" would have to perform a spinning motion around each other in order to become separated. If the daughters are to be formed side by side and free, the parent portion must rotate around its axis. Immediate separation of daughter duplexes not only is required for one replication cycle following another without delay, but also is evident from the autoradiographs obtained by CAIRNS (Fig. 5). Obviously the fork at which physical separation of the daughter molecules occurs is located at the point of replication. Thus unwinding may take place at the parent or unreplicated side of the fork. CAIRNS (1963 b) tentatively assumed that the unreplicated part of the molecule rotates in its entirety. A "swivel point" or point of free rotation (possibly a single chemical bond) was assumed to exist at or near the origin of replication (Fig. 7).

At first sight this appears satisfactory, but some quantitative aspects have to be considered also. First, the speed of rotation that has to be maintained in order to unwind the entire helix can be considerable and for E. coli and a generation time of 30 minutes would amount to 10,000 revolutions per minute. Second, some internal friction is likely to exist and to interfere with such rapid rotational movement within the DNA-plasm. Such friction is caused by the binding of part of the interstitial water by the DNA itself, mainly due to ionization of the phosphate groups. Both degree of ionization and interhelix distances will determine the degree of formation of water shells around the DNA helices. The DNA-plasm contains about 1.5 per cent DNA (weight/volume). In a maximum degree of ionization (as in sodium deoxyribonucleate) such solution of DNA forms a solid gel. Actually the degree of ionization of intracellular DNA will be lower due to the presence of more strongly bound ligands such as calcium, magnesium, and polyamines (see also Section I, J. p. 50), consequently interhelix forces will be weaker. Third, the genophor is about thousand times longer than the space in which it is contained, and therefore must be wound or folded up to a considerable extent to form the accumulation of DNA we referred to above. It is difficult to see how such a structure in all its parts could withstand transmission of the torque necessary to make it rotate in its entire length at such considerable speed.

In view of our ignorance concerning the chemical status of intracellular DNA it is difficult to be more precise. In an attempt towards a quantitative formulation, LEVINTHAL and CRANE (1956) calculated the energy which is necessary to overcome the viscous drag during uncoiling of a DNA helix. Their assumptions regarding the number of turns to be unwound and the velocity of rotation are based on actual *E. coli* data, but the resulting estimate undoubtedly is low, since the effective radius of the helix was assumed to equal the radius of the dehydrated duplex molecule and the viscosity of the "solvent" was assumed to equal that of pure water. Both assumptions are unrealistic.

ATKINS (1964) states that if the concept of semiconservative replication is adhered to and if better explanations on topological and dimensional grounds prove impossible, "further consideration should be given to the possibility of a disruptive mechanism".

"Weak spots" and one-stranded "adhesive sites" were demonstrated in bacteriophage DNA (RIS and CHANDLER 1963, HERSHEY et al. 1963, DAVIDSON et al. 1964, RHOADES et al. 1965, FREIFELDER and KLEINSCHMIDT 1965, KAISER 1965, HERSHEY and BURGI 1965, ABELSON and THOMAS 1966), but similar evidence thus far has not been found in bacteria (see KLEINSCHMIDT et al. 1961, LANG and KLEINSCHMIDT 1964, LANG et al. 1964). The presence of a small amount of amino acids possibly associated with DNA has been taken as an indication of some chemical heterogeneity of the genophor (BENDICH and ROSENKRANZ 1963), and some isolates of bacterial DNA have been termed subunits of the genophor (MASSIE and ZIMM 1965), but suggestions on these grounds with reference to the problem of uncoiling are highly speculative.

A way to resolve this difficulty is indicated in recent observations on the mechanism of thymineless death and on the repair of damage caused by irradiation. Thymineless death is a condition which occurs in thymineless mutants upon withdrawal of the external supply of this essential DNA precursor (COHEN and BARNER 1954, 1955). The condition exists during periods of DNA replication as well as during periods of DNA transcription (synthesis of messenger RNA). In the absence of either process the cells are immune against thymine withdrawal (MAALØE and HANAWALT 1961, GALLANT 1962, GALLANT and SUSKIND 1962, HANAWALT 1963, ROLFE 1967). PAULING and HANAWALT (1965) found that during DNA replication following a period of thymine starvation short reaches of DNA originated from a non-conservative type of DNA replication (i.e., they contained newly introduced nucleotides in either strand). They concluded that thymine starvation prevents repair of single-strand breaks produced by such starvation or by other causes or by both. Single-strand breaks are known to be produced by the decay of incorporated ^{32}P and by X-rays and are also produced during repair of damage by ultraviolet light. These agents resemble the thymineless condition in that damage is more severe during periods of

DNA transcription. In combination with thymine starvation viability is affected in a synergistic fashion (ROBERTS 1960, BILLEN 1963, RASMUSSEN and PAINTER 1963, GINSBERG and JAGGAR 1965 a, b).

PAULING and HANAWALT (1965) and HANAWALT (1966) suggest that transcription of DNA normally may involve the formation of single-strand breaks. Such breaks may occur at each operon and thus may provide points of free rotation for the unwinding of a small segment of the DNA as required for replication. The integrity of the DNA duplex would be restored by a repair mechanism requiring thymine. Similar breaks would occur during transcription as indicated by the effect of thymine starvation on messenger synthesis (LUZZATTI 1966, for detail see Section I, H, 3, p. 43).

These observations may shed new light on the process of DNA synthesis in the absence of replication and on various questions in relation to the dependence of mutation frequency and expression on replication and transcription of DNA (WITKIN 1958, RYAN and KIRITANI 1959, NAKADA 1960, NAKADA et al. 1960, NAKADA and RYAN 1961, KUBITSCHEK 1966).

The synthesis of new strands requires the existing strands to act as templates. This is assumed to involve exposure of the purine and pyrimidine bases normally situated in the interior part of the helix and chemically inaccessible due to hydrogen bonding. Some speculations are concerned with the possible template function of the two grooves or furrows which exist alongside the helix and in which a specific steric pattern is created by the sequence of the bases. Other models (e.g. by STONEHILL 1965) assume exposure of the bases by a swiveling motion around the phosphodiester bond. Results by several authors indicate that temporary and localized loosening of the duplex structure may occur in order to permit copying of the base sequence. Evidence for the existence of single-stranded DNA at the replication point was obtained by GOLDSTEIN and BROWN (1961), ROLFE (1963), ROSENBERG and CAVALIERI (1964) and KIDSON (1966). SETLOW (1960), SETLOW and SETLOW (1960) and SETLOW and BOYCE (1961) found UV action spectra during bacteriophage replication to be characteristic of single-stranded DNA.

F. Regulation of DNA Synthesis

1. Timing of DNA Synthesis in Relation to the Cell Cycle and the Influence of the Metabolic State on DNA Synthesis

DNA synthesis in rapidly growing bacteria was found to be a practically continuous process. This was established in experiments using pulse-labeling with ^{3}H-thymidine and ^{32}P (SCHAECHTER et al. 1959, McFALL and STENT 1959, PACHLER et al. 1965), and by DNA determination in germinating spores and in *E. coli* populations synchronized by filtration (YOUNG and FITZ-JAMES 1959, CUMMINGS 1965). The accuracy of these determinations was sufficient to reveal arrests of DNA synthesis as short as 3 per cent of the generation time. From the number, in a growing population, of cells resistant against thymineless death, the fraction at the point between rounds of replication was estimated at 3 per cent. A similar conclusion was reached

by Cairns (1963 a) who found the length of pulse-labeled DNA to be related
to the length of the genophor in the same way as was the duration of the
pulse to generation time. From the results of Young and Fitz-James and of
Cairns it is also evident that DNA replication within an individual cycle
proceeds linear with time.

When growth is slower due to suboptimal nutrition, DNA synthesis
may be slowed down in three possible ways: alternate replication of nucleoids, reduction of rate of synthesis, and intermittent synthesis.

E. coli 15 T⁻met⁻try⁻ in mineral salts-glucose medium grows with a
doubling time of 40 minutes. The cells contain two nucleoids each, which
are replicating continuously and simultaneously. In succinate medium,
doubling time is increased to 70 minutes; the rate of DNA synthesis of the
population is reduced accordingly. Experiments involving arrest and
reinitiation of genophor replication and double-labeling reveal, however,
that replication of single nucleoids proceeds at the same rate in either
medium, the reduction in overall rate of synthesis being due to the fact
that at any one time only 50 per cent of the nucleoids are in a state of
replication. According to these experiments, succinate-grown cells still
contain two nucleoids which, however, are replicated alternately and in
sequence (K. G. Lark and C. Lark 1965).

At still lower growth rates as found in E. coli growing on aspartate at
a doubling time of 120 minutes, the rate of DNA synthesis in individual
nucleoids was affected. The cells contained one nucleoid each which was
in the process of replication throughout the cell cycle (C. Lark 1966).

If more than two hours are required for one doubling, DNA synthesis
becomes discontinuous. E. coli 15 T⁻ grew on proline or acetate with doubling times of 180 and 270 minutes respectively. Its single nucleoid was replicated during the second half of the cell cycle only (C. Lark 1966, see
also Ecker and Schaechter 1963 b). In ³H-thymidine labeling experiments
Schaechter (1961 b) and Maaløe (1961, 1963) found that at doubling times
of 50 minutes and longer at 37° C the fraction of cells in a population that
failed to incorporate a short pulse of labeled precursor began to become
noticeable. Rosenberg et al. (1967) observed discontinuous synthesis of
DNA in Myxococcus xanthus growing at rate of one doubling per
270 minutes.

In new approaches to this problem, the relative timing of DNA replication and cell division was elucidated to a considerable extent. Kubitschek
et al. (1967) separated chemostat-grown cells by volume in a sucrose gradient
and measured DNA synthesis by the incorporation of ³H-thymidine. During
unrestricted growth all but the smallest cells (approximately 4 per cent)
synthesized DNA. At doubling times longer than two hours onset of DNA
synthesis was delayed. At generation times of 8 to 12 hours, DNA synthesis
started in cells the volume of which was 1.5 to 1.7 times that of cells immediately after division. Indirect evidence was obtained for the termina-
tion of DNA synthesis some time before cell division.

Helmstetter (1967), Helmstetter and Cooper (1967) and Cooper and
Helmstetter (1967) applied the principle of membrane-binding and elution

(HELMSTETTER and CUMMINGS 1964) to the analysis of the pattern of DNA replication in cells of *E. coli* B/r labeled with ³II-thymidine. The growth medium was modified to produce doubling times between 20 and 60 minutes. The rates of DNA synthesis in fractions eluted at various times after labeling are in agreement with a model according to which, irrespective of growth rate, replication of the genophor is accomplished in 41 minutes, and the time between the end of a round of replication and cell division is 22 minutes. Cells growing slower than 40 minutes per doubling have "gaps" between rounds of DNA synthesis; cells doubling in less than 40 minutes show periods of rapid synthesis of DNA which are explained in terms of multiple replication points on single genophors ("multifork replication").

CLARK and MAALØE (1967) performed pulse-labeling experiments with filter-eluted *E. coli* B/r cells. They conclude that replication starts 20 to 30 minutes before cell division. Doubling times in glucose, glycerol, and succinate media were 30, 45, and 60 minutes respectively. The observation that succinate-grown cultures replicate their genomes in an alternate fashion (K. G. LARK and C. LARK 1965) could not be confirmed with this strain.

The cell cycle of *B. subtilis* is characterized by "gaps" in the process of DNA synthesis when cells are growing on acetate and succinate at generation times of 200 and 500 minutes respectively while synthesis is continuous in cells growing on glucose at a doubling time of 80 minutes (EBERLE and LARK 1967).

These investigations have largely superseded the results of early attempts to detect timing of nucleoid replication in living or stained bacteria with the light microscope (KNÖLL and ZAPF 1952, 1954, LARK et al. 1955, PERRET 1958), but this is by no means true for the cytology of nucleoid division which is still incompletely understood. Unfortunately the sophistication of the experiments described above has not been achieved in cytological investigations. Only the short report by PERRET (1958) deserves special mentioning in that it partly anticipates results presented in the preceding paragraphs.

Electron micrographs show that nucleoid division in certain instances is barely completed when daughter cells are separated by the ingrowing cell wall (Fig. 62, p. 122) (CHAPMAN 1959, 1960, CONTI and GETTNER 1962, GRULA and SMITH 1963). Such observations can be made with the light microscope also and caused earlier workers to speculate on the "passive constriction and separation of nuclear material". Since then, the mass equivalent of the bacterial genome has been determined, and simple considerations of DNA content and DNA economy of bacteria render the idea of passive or random distribution untenable. The observed cytological patterns and forms of coordination of DNA replication, nucleoid division, and cell division require that replication of any segment of the genophore is immediately followed by separation of the daughter duplex structures and by their incorporation into nuclear entities (nucleoids) which can become physically separated without difficulty or delay. If actual separation does not occur until shortly before cell division, the daughter structures involved nevertheless are preformed and exist as already largely independent nuclear entities.

In this manner, observations on the timing of synthetic and divisional processes represent cytologically relevant information and can exert certain constraints on the interpretation of nucleoid fine structure.

2. Mechanisms of Regulation

a) Protein Requirements

There is little indication that precursor synthesis and availability of polymerase are of general importance in the regulation of DNA synthesis. A variety of conditions causing inhibition of DNA synthesis did not interfere with precursor availability nor with the operation of the synthetic enzymes (BILLEN 1960a, DOUDNEY and BILLEN 1961, LARK 1961, SCHAECHTER 1961a, KOHIYAMA et al. 1963). During thymine starvation various precursors accumulate in different patterns (NEUHARD and MUNCH-PETERSEN 1966), and experiments by OKAZAKI et al. (1959) indicate that thymine may have other functions besides being a DNA precursor (see LARK 1963 for review).

Inhibition of protein synthesis, however, profoundly affects the pattern of DNA synthesis. Starvation for essential amino acids of appropriate mutants, sometimes also chloromycetin treatment, interferes with continued synthesis of DNA and prevents resumption of synthesis after its arrest by irradiation or chemical action (PARDEE and PRESTIDGE 1956, BARNER and COHEN 1957, GROS and GROS 1958, HAROLD and ZIPORIN 1958, 1959, DRAKULIĆ and ERRERA 1959, BILLEN 1959a, b, 1960b, 1961, DOUDNEY 1961b).

If a population of E. coli 15 T−A−U− (incapable of synthesizing thymine, arginine, and uracil) is brought into a medium devoid of arginine and uracil, the fraction of cells immune against thymine withdrawal (see p. 20) will increase from 3 per cent to 100 per cent. This means that the cells of that population are developing towards a physiological state normally shared by a few members of the population only. During the experiment, the DNA content of the culture increases by 40 per cent. This figure suggests that cells in various phases of their DNA replication cycle may complete the round of replication just under way without being able to initiate a new cycle (MAALØE and HANAWALT 1961). This conclusion was verified in autoradiographic experiments. The fraction of the population capable of incorporating a short pulse of label decreased from 100 to 0 per cent in the course of the experiment indicating an arrest of DNA synthesis in an increasing number of cells (HANAWALT et al. 1961).

Alignment in this manner of DNA replication does not result in synchronization of replication and division cycles. This is due to the fact that DNA synthesized during amino acid starvation is abnormal in that, during subsequent cycles, it is replicated at a lower rate (LARK et al. 1963). The addition of amino acids restores the ability of the cells to synthesize DNA even during periods when no DNA synthesis takes place (as in the absence of thymine, K. G. LARK 1966). This effect is believed to be of more general importance in that the rate of protein synthesis during any one round of

DNA synthesis may determine the rate of DNA synthesis in the subsequent round (K. G. LARK 1966). In any case, these experiments demonstrate the requirement for initiation of DNA synthesis of a protein not immediately and permanently associated with the DNA.

If two successive experiments involving "alignment" of replication cycles and subsequent re-initiation of replication by addition of amino acids are performed on a population of *E. coli* T⁻A⁻U⁻ cells, and two kinds of label are introduced into the DNA for short periods after reinitiation of synthesis by addition of ³H-thymidine and 5-bromouracil respectively, both kinds of label are incorporated into identical segments of the genophor. This indicates that the starting point of replication in subsequent cycles remains the same (LARK et al. 1963, PRITCHARD and LARK 1964).

In the early experiments no distinction could be made as to whether the synthesis of protein or of RNA was the critical step necessary for continued synthesis of DNA. The similarity of results with amino acid starvation alone as well as earlier observations by GOLDSTEIN et al. (1959), NAKADA (1960) and OKAZAKI and OKAZAKI (1959) points towards a specific requirement for protein rather than RNA synthesis.

In *E. coli*, chloromycetin (chloramphenicol) was not quite

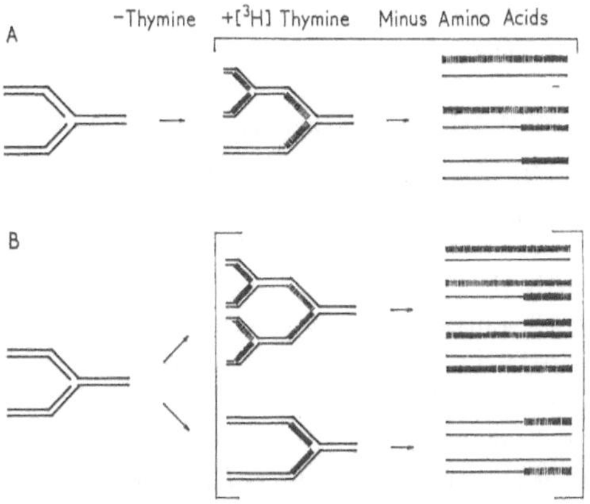

Fig. 8. Two models for the replication of the *E. coli* genophor following thymine starvation. Replication in the absence of required amino acid is shown. (*A*), A premature replication cycle is initiated in one of the two available partial replicas of each genophor. (*B*), In half of the chromosome population, a premature cycle is initiated in both of the available partial replicas of each chromosome. No premature replication cycle is initiated in the other chromosomes of the population. — The models are based on the quantitative interpretation of the data of PRITCHARD and LARK (1964). The number of conserved units, labeled during replication, are shown (——+·+·+——). — Reproduced from LARK and BIRD (1965 b).

as effective in preventing initiation of new rounds of DNA synthesis as was amino acid starvation. This seemed to indicate that amino acids did not exclusively act as protein precursors. But a direct interference with DNA synthesis of amino acids (similar to amino acid-controlled RNA synthesis, KELLENBERGER et al. 1962) could be ruled out by FRIESEN and and MAALØE (1965).

Meanwhile it has become apparent that a twofold involvement of amino acids in the regulation of DNA synthesis exists, that both mechanisms involve synthesis of a protein, but that only one of them is chloromycetin-sensitive. This was established in the following manner: Thymine withdrawal in thymineless *E. coli* can result in the premature initiation of a new replication cycle in one of the daughter structures ("multifork replication", Fig. 8, PRITCHARD and LARK 1964, BILLEN 1964, LARK and BIRD 1965 b, K. G. LARK 1966). This effect which leads to the formation of three daughter nucleoids is chloromycetin-sensitive and thus is caused by the availability of a surplus protein which may accumulate during periods of arrest of both DNA synthesis and cell multiplication (NAKADA 1960, C. LARK and K. G. LARK 1964).

Contrary to this effect, the "normal" initiation of DNA replication is amino acid-dependent without being chloromycetin-sensitive (K. G. LARK 1966). The presumably proteinaceous factor synthesized for "normal" initiation does not accumulate in the absence of DNA synthesis (i.e. during periods of thymine starvation). Therefore it either is a part of the DNA structure (PRITCHARD and LARK 1964, C. LARK and K. G. LARK 1964) or its formation is dependent upon the completion of the DNA structure. This compound may be a product of ribosome-independent protein synthesis, a process for which a few examples are known (see BROCK 1961, C. LARK and K. G. LARK 1964 for references). The term "normal" in this section is used with some reservation since multiple replication points originating from premature initiation may not be as uncommon as initially assumed (see p. 23).

Phenethyl alcohol and canavanine (an analogue of arginine) interfere with the initiation of new cycles of DNA replication in a similar way as does amino acid withdrawal. In the case of canavanine, inhibition of synthesis of an essential protein is a likely explanation (TREICK and KONETZKA 1964, K. G. LARK and C. LARK, cit. by K. G. LARK 1966, SCHACHTELE 1965).

In contrast to *E. coli*, synthesis of normal initiator in *B. subtilis* is chloromycetin-sensitive (YOSHIKAWA 1965), and multifork replication involving b o t h daughter structures ("dichotomous replication") is found as a normal event during rapid growth (YOSHIKAWA et al. 1964, OISHI et al. 1964). Completion of DNA replication cycles upon withdrawal of an essential amino acid as well as induction of an extra cycle and inhibition of this latter event by chloromycetin was also observed in *Lactobacillus acidophilus* strain R-26 (SOŠKA and LARK 1966).

Independent evidence of protein participation in the initiation of DNA replication was obtained from experiments with a temperature-sensitive strain of *E. coli*. This mutant was unable to start new rounds of replication at 40° C whereas it multiplied normally at 30° C (KOHIYAMA et al. 1963). The effect of chloromycetin resembled that of incubation at 40° C. The DNA polymerase system was found unaffected by the mutation.

After treatment with X-rays, resumption of DNA synthesis is amino acid-dependent and chloromycetin-sensitive (BILLEN and HEWITT 1965). During recovery from irradiation with X or UV rays DNA synthesis was resumed in a segment of the genophor different from the one replicated immediately before radiation damage was incurred. Whether the new starting point is identical with the normal origin of replication is not clear (BILLEN et al., 1965, HEWITT and BILLEN 1965).

GOGOL and ROSENBERG (1964) found primer activity for DNA polymerase of *Myxococcus xanthus* to be reduced 90 per cent, if such DNA was isolated from stationary phase rather than log phase cells. This might indicate the presence of an unknown factor reducing polymerase activity.

b) Models of Regulatory Mechanisms

The two factors introduced in the preceding section and presumably of polypeptide nature, currently are holding key positions in considerations concerning the mechanism of regulation of DNA synthesis in bacteria. When attempting to become familiar with such hypotheses, one should, however, bear in mind that even concepts such as a protein molecule attached to DNA and replicated with it are purely speculative at the present time. For the purpose of this review, models will be considered insofar as they explain experimentally established facts with an absolute minimum of additional assumptions. Their discussion will prove helpful for those who realize their heuristic value but nevertheless resist the attempt of having their imagination compelled onto an unnecessarily narrow line of thought.

First we shall discuss the term r e p l i c o n as introduced by JACOB and BRENNER (1963). The term replicon or the synonymous designation "a u t o n o m o u s   u n i t   o f   r e p l i c a t i o n" refers to what has been termed earlier "organized DNA molecule". It designates a DNA molecule capable of replicating under its own control. The concept originates from considerations on the co-existence of genetical elements within a single cell and their patterns of physical association and mutual control, and from the observation that fragments of DNA are incapable of replication unless associated with another DNA molecule which is of the organized or replicon type. Accordingly, a replicon consists of DNA plus functional groups or elements physically associated with it. Conversely, such functional groups by definition act only on the DNA element or genophor to which they are physically attached. Replication initiated in such unit will, conditions permitting, proceed along the entire structure. The location of the functional groups eventually will designate the starting point of replication; their attachment to one of the DNA strands may determine the polarity of replication, particularly in replicons of circular shape.

Obviously the entire genome of *E. coli* is organized as a single replicon, and so are the genomes of phage and specialized genetic elements, the episomes. If one of the latter elements becomes integrated into a bacterial genome (by recombination), its replication will be subject to the control of the functional group that is part of the bacterial genome. Conversely, a segment of bacterial DNA, if integrated into an episome or phage genome, will replicate under the control of the latter.

The functional group which controls the replication of the DNA segment or replicon to which it is attached was named r e p l i c a t o r (JACOB and BRENNER 1963). This term is chosen in analogy with the term "operator", designating a functional group which controls messenger synthesis along a unit of transcription or operon (JACOB and MONOD 1961). The replicator is the site of action of a hypothetical protein, the i n i t i a t o r. Interaction of both replicator and initiator is required for replication of the DNA replicative unit. The replicon concept in this point differs from the operon concept. The operon is negatively controlled by a hypothetical molecule, the repressor. Removal or inactivation of repressor starts transcription.

While the replicator is an integral part of the DNA structure which it controls, the initiator is mobile but specific. Within the boundaries of a cell, it may trigger replication in any replicon of a specific kind (bacterial, phage, episomal). To explain such specificity it is assumed that the initiator is a protein derived from a special operon which is part of the genetic message of the kind of replicon controlled.

The idea that the initiator is an enzyme which travels with the replication fork has been abandoned, since the concentration of enzymes involved in DNA synthesis is little affected under conditions when initiator synthesis is inhibited (BILLEN 1962, KOHIYAMA et al. 1963).

In the preceding section, the existence of a protein was described which is synthesized via a chloromycetin-sensitive mechanism and which is essential for the initiation of DNA replication. This protein fulfills the requirements of the initiator as conceived by JACOB and BRENNER (1963).

The mode of action of the initiator in the light of experimental evidence is discussed in considerable detail by K. G. LARK (1966). It is implied that the initiator is a rather stable molecule that may perform its functions for more than one replication cycle. Initiation of the cycle requires attachment of the initiator to the replicator site. During the cycle, the initiator remains attached. This accounts for the (normally) impossible initiation of a second cycle before the first is completed. A signal generated at completion of the cycle liberates the initiator and renders it available for another round.

During rapid growth, each of the two daughter genophors emerging from a round of replication will immediately embark in a new replication cycle. This means that the cell, during the first cycle, must provide one more molecule of initiator. Failure to do so will result in the failure of one of the daughter structures to replicate, and restriction of replication to the other will result in an apparent linear increase of DNA content of the population. Conversely, if an excess of initiator is produced, this may contribute towards the premature initiation of new cycles. Such accumulation of initiator molecules may occur during periods of thymine starvation.

According to this model, inhibition of initiator formation causes DNA synthesis to proceed in a linear fashion. It does not result in a complete cessation of DNA synthesis, since the liberated initiator is available for re-attachment. The observed patterns of discontinuous synthesis therefore require modification of the model. It is proposed that chloromycetin-sensitive protein synthesis (a growth rate-dependent process) furnishes a second protein referred to as "proreplicator". The proreplicator would provide an attachment site for one of the daughter structures, such attachment being mediated by the initiator. Upon completion of the cycle, the observed chloromycetin-independent mechanism would serve to convert the proreplicator site into a replicator and to liberate the initiator which latter then would become available for new initiation, provided a new proreplicator site has been synthesized. The proposed mechanism is illustrated in Fig. 9.

The model prompts a few questions: First, the model is compatible with premature initiation in one of the daughter structures as pointed out by the author. Premature initiation in both strands (dichotomous replication) is more difficult to accommodate. Second, the model requires dislocation of the origin of one of the daughter structures from the origin, possibly to some membrane site. During replication, such membrane would represent the only mechanical connection between the old origin (replicator) and the new (proreplicator) site. Such physical separation is at a variance with the autoradiographs by CAIRNS (Fig. 5). One may argue that the small distance

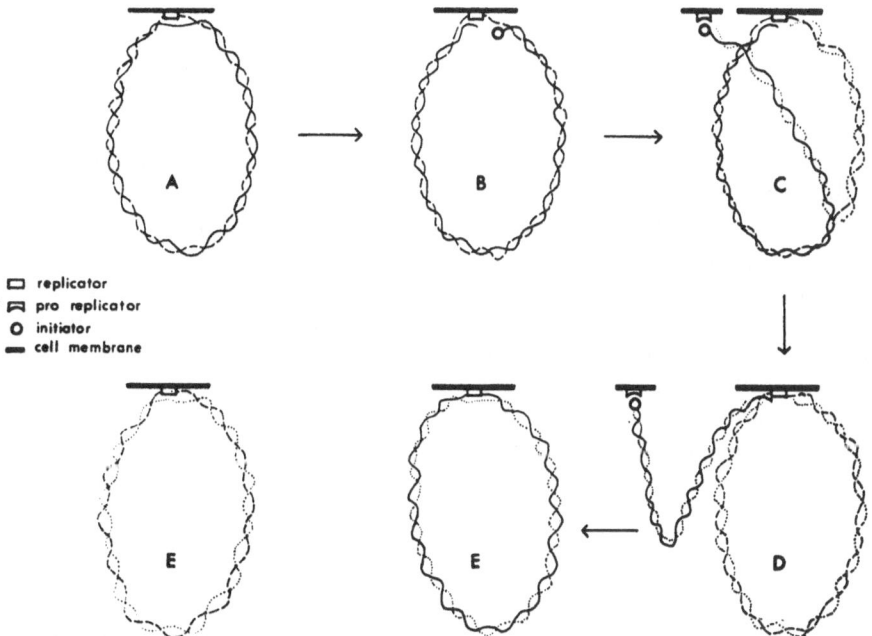

Fig. 9. Replication of a genophor in *Escherichia coli*, diagrammatic representation. One strand of the genophor is permanently attached to the cell membrane by a protein, the replicator. The other is detached by interaction with the initiator, and is temporarily attached to the cell membrane via a protein, the proreplicator. Permanent attachment of this strand results when replication is completed, thus converting the proreplicator into a replicator. — Reprinted from K. G. LARK (1966).

between the two sites is not revealed in these l i g h t micrographs. On the other hand, any physical link consisting of protein and cytoplasmic membrane material would be among the structures most heavily affected by the isolation technique, and its persistence throughout the procedure is improbable. Mechanical separation of the old and the new replicator at an early phase of the cycle also requires one of the daughter structures to develop a "homing instinct" of its posterior end in order to accomplish ring closure upon completion of the replication cycle proper. The nature of such mechanism is not clear. Finally, it is proposed, that in every cycle and in one of the daughter structures formed, the replicator becomes attached to the opposite polynucleotide strand. This idea was conceived in order to explain the segregation pattern in succinate-grown *E. coli*. While this type of segregation will be discussed in detail in the next section, it is felt that in proposing this mechanism K. G. LARK abandons the most plausible explanation for the maintenance of polarity during genophor replication, i.e. the attachment of the replicator to a specific polynucleotide strand, the polarity of which then would determine the polarity of replication.

An alternate mode of regulation, particularly concerning the requirement for proreplicator sites in the sense indicated above, has been proposed by MAALØE (1964): According to MUNCH-PETERSEN and NEUHARD (1964) and NEUHARD and MUNCH-PETERSEN (1966) nucleotides, especially deoxyadenosinetriphosphate (dATP) accumulate during thymine starvation of *E. coli* cells. Such accumulation does not occur when cells have been permitted to complete DNA replication cycles (or: when immune against thymineless death). This would indicate the existence of a structure with enzyme functions which is required for DNA replication and which disintegrates when the replication cycle has been completed. This would provide a simple positive type of regulation by protein synthesis of the initiation of new rounds of replication.

The question of membrane attachment of the nucleoid is discussed in greater detail in the sections on segregation and fine structure of nucleoids. The aspects of possible membrane attachment with respect to replication, however, must be considered in this section. Unfortunately, little discrimination is made in most discussions as to whether the cytoplasmic membrane, the cell wall mucopeptide or both are the sites of attachment. The chloromycetin-independent mode of formation or completion of new replicator was taken to indicate association with the rigid mucopolymer layer, since the uptake of amino acids into the mucopolymer (mucopeptide) is known to be chloromycetin-insensitive (see C. LARK and K. G. LARK 1964 for references), and the same holds for the synthesis of ribosomal and certain enzyme proteins. In drawing conclusions from these observations, one should start with the determination of the structural element common to these synthetic processes. This turns out to be not the membrane but rather the DNA as the established site of the coded information, and unless recourse is taken to entirely unknown factors, such reasoning would only lead to the conclusion that the replicator is synthesized at the DNA or at a template site derived from the DNA. If the analogy between mucopeptide and replicator synthesis is taken to indicate a spatial relationship between both structures, one would have to reason that the cell wall may be the template site for both mucopeptide and replicator synthesis—a somewhat arbitrary interpretation.

An entirely different model of genophor replication was proposed by JACOB et al. (1963, 1966). According to these authors, the replication fork and the enzymatic machinery involved in DNA synthesis (DNA polymerase) may be located at the cytoplasmic membrane where they form a "c o m p l e x o f r e p l i c a t i o n". The replicating genophor is propelled along or through this complex. This model receives its original support from observations on the replication of the F episome in F+ cells of *E. coli* K 12 and F-controlled replication of the bacterial genome in Hfr cells. In the latter case, one of the daughter structures is injected into the recipient cell (see p. 41 for details). The proposal that the n o r m a l bacterial "complex of replication" is also membrane-bound, is a generalization from these observations. Nevertheless, the model is supported by observations which link the replication fork to a membrane site (GOLDSTEIN and BROWN 1961, HANAWALT and RAY 1964, PETTIJOHN and HANAWALT 1964, SMITH and HANAWALT 1965, GANESAN and LEDERBERG 1965 a, b). The idea of a genophor passing along a fixed point may also aid in understanding the integration of DNA fragments into specific segments of the genophor ("homing instinct" of prophage and episomes, W. BRAUN 1965), provided that such integration requires DNA replication.

The difficulties presented by the model are that of a massive accumulation of precursors at the membrane site which is required to permit rapid DNA replication, and the question of coordination and cooperation of multiple complexes of replication located on the same genophor. This concerns the cooperation of bacterium-controlled replication and replication at conjugation in *E. coli* Hfr, premature initiation of replication cycles (multifork replication), the occasional relocation of the origin of replication in *E. coli* and *Salmonella typhimurium* (CHAN and LARK 1967 and CHAN, personal communication).

The cytological aspects, consequences and possibilities of these models will be discussed in the Section on nucleoid fine structure, their possibilities for the understanding of genome segregation and the regulation of cell division will be evaluated in the two subsequent sections.

G. Genome Segregation

A distinct pattern of genome segregation was first predicted from the delay in appearance of recessive mutants in conjunction with microscopic observations of the number of nuclear bodies per cell (WITKIN 1951) and subsequently was examined with the aid of labeled DNA precursors such as ^{3}H-thymidine and ^{32}P-phosphate (FORRO and WERTHEIMER 1960, HERMAN and FORRO 1964, FORRO 1965, STENT and FUERST 1960) and by experiments on the segregation pattern of recombinant linkage groups following conjugation (TOMIZAWA 1960). These results are in accordance with the diagram shown in Fig. 10 according to which, upon DNA replication, two nucleoids are formed in adjacent positions containing one double-stranded unit each. The nucleoids are distributed into daughter cells according to their intracellular arrangement.

A different segregation pattern was observed in succinate-grown cells of *E. coli* (LARK and BIRD 1965 a). Such cells contain two genetic complements each which are replicated alternating and in sequence (K. G. LARK and C. LARK 1965). At cell division, each genetic complement contributes half of its progeny to each of the daughter cells. FORRO (1965) observed a similar pattern and suggested that two circular structures may have combined by a recombinatory event to form a single replicating unit of double length. The increase in DNA content after amino acid withdrawal, however, indicates the completion of a replication cycle in a DNA unit of normal size. These and other observations show that in succinate-grown *E. coli* both

DNA complements still are under individual control with respect to replication and do not operate as a single replicon. Some coordination must. however, exist to ensure their alternate mode of replication.

The latter segregation pattern resembles that of JACOB et al. (1963, 1966) for the joint segregation of nucleoids and sex factors (see Fig. 11). As expressed by EBERLE and LARK (1966), such mechanism ensures that templates that have been used n-times are separated from those that have been used (n+1)-times. The pattern produces virtual diploidy (LARK), which more exactly (see Introduction) should be named di-energidy. This according to far-reaching speculations by JACOB et al. (1966) may not be accidental or

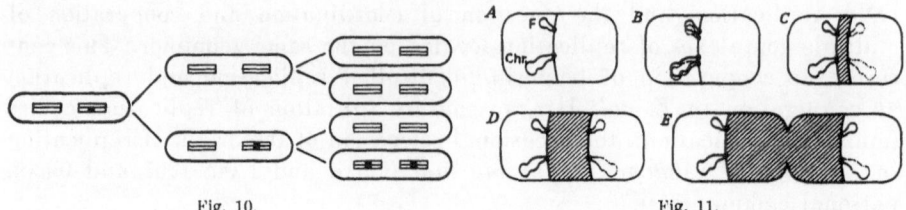

Fig. 10.   Fig. 11.

Fig. 10. Normal segregation pattern in a growing population of bacteria containing two nucleoids per cell. Diagram based on data by WITKIN (1951), TOMIZAWA (1960), LARK and BIRD (1965 a, b, for glucose-grown *E. coli*), and others.

Fig. 11. "A model for the regulation of DNA synthesis and for an equal distribution of DNA among the daughter bacteria. The bacterium (F^+) carries two independent self replicating units: a bacterial genome and an independent sex factor. The two replicons are assumed to be independently attached to specific sites at the bacterial surface along the equatorial perimeter. At a certain stage of the bacterial cycle of multiplication, the cell surface is assumed to transmit to the replicons the signal initiating their replication, which proceeds linearly along the two structures. For each of the replicons, two daughter structures are formed which are assumed to be attached side by side to the bacterial membrane. Elements of the bacterial membrane are assumed to grow between the two planes of attachment of the daughter replicons, putting them progressively apart. During this process, no further replication is permitted, until the growth of the bacterial membrane has reached a certain point and a signal allowing a new replication cycle is given. The process is oversimplified for representation in the sense that (1) bacteria have generally 2 to 4 (and not 1 to 2) DNA complements per cell, DNA replication being one cycle ahead of cell division and that (2) every step is assumed to be completed before the next one can be initiated. The important point of the model is a sequence: initiation of DNA synthesis—growth of the bacterial membrane between the attachment points—initiation of a second cycle of DNA synthesis etc. Septum formation and separation of daughter bacteria is not assumed to be a necessary step of the sequence but to be correlated with some step since many treatments are known to result in the formation of "snakes" in which bacteria become elongated and synthesize DNA without dividing. In this model, the unit of segregation is formed by the equatorial perimeter, to which the cellular replicons are assumed to be attached." — Reprinted from JACOB et al. (1963).

unique to bacteria but may represent a mechanism homologous to the more complex ones found in eucaryotic cells.

Claims by EBERLE and LARK (1966) that genome segregation in *Bacillus subtilis* follows an identical pattern have been challenged by RYTER and JACOB (1966b, 1967) who found a random distribution of daughter genomes (Figs. 12 and 13).

The involvement of the cell membrane or cell wall in nucleoid division and segregation is considered likely by most authors but as yet is incompletely understood. It is suggested that nucleoids grow apart as the result of a specific growth pattern of certain membrane layers. Unfortunately, the growth pattern of these layers is incompletely known. The patterns observed by immunofluorescence techniques are those of the peripheral, antigenic layers which may or may not be identical with the rigid mucopolymer layer. The nucleoid may be attached to the innermost layer only, i.e. the cytoplasmic membrane, and the structure of plasmalemmosomes seems to suggest that the growth pattern of that layer is not identical with that of the mucopeptide layer. Tellurite labeling of the membrane (RYTER 1967) seems to offer a definite advantage over the older techniques. In this way RYTER found that a growth pattern of the membrane exists in *Bacillus subtilis* which is in perfect accordance with that shown in Fig. 11.

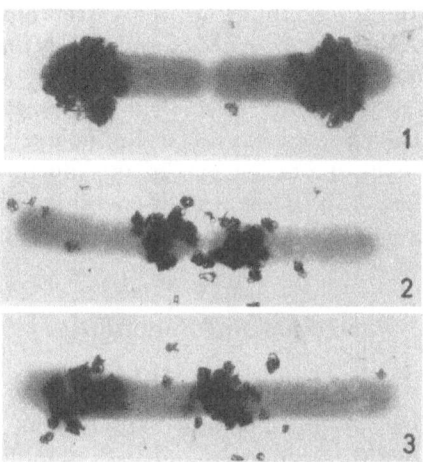

Fig. 12. Autoradiographs of cells grown out of spores of *Bacillus subtilis*, labeled with ³H-thymidine and germinated in cold medium. After the second replication, three types of distribution of label are obtained as shown. Frequencies are: types one and two, 25 per cent each, type three: 50 per cent. Numbers of figures correspond to those in the diagram (Fig. 13). — Reprinted from RYTER and JACOB (1966 b).

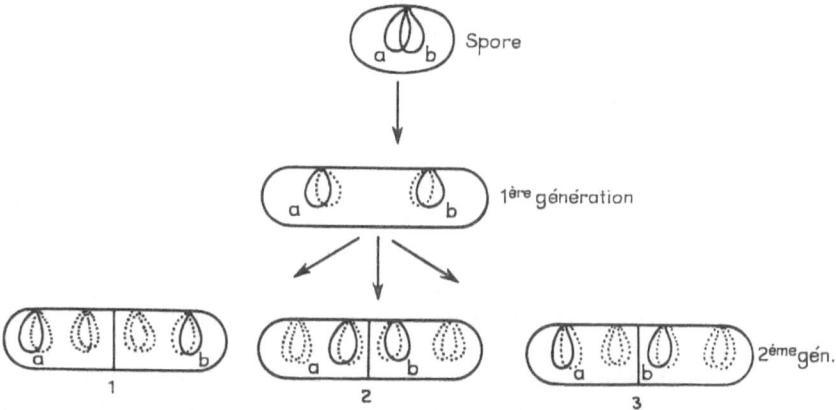

Fig. 13. Diagrammatic representation of genome segregation in *Bacillus subtilis*. Three hypotheses are shown: (1) The original conserved units are found at the poles of the chain. — (2) The original conserved units are found at the inner poles of the cells in the chain. — (3) The units are distributed at random. In this case, 25 per cent of pairs show an arrangement of the type 1, 25 per cent of type 2, and 50 per cent of type 3. The last model is supported by observations. — Reprinted from RYTER and JACOB (1966 b).

The principal disadvantage of the model (Fig. 11) is that it represents an attempt to explain too much at the same time. The authors admit that DNA replication and wall growth in reality occur simultaneously. This and the established wall growth pattern plus the fact that the number of nucleoids is restricted to four per cell impose constraints on the model which interfere with the proposed segregation pattern (mimicking diploidy) or with the permanent attachment of the replication fork or both. As will be discussed later, the dilemma can be resolved if it is assumed that a new point of wall attachment is formed in every nucleoid in an early replicated region (i.e. a region different from the replication fork) (RYTER 1967, 1968, see p. 127).

A segregation pattern resembling diploidy was observed in giant cells of *Pasteurella pestis, Micrococcus aureus* and *Escherichia coli* formed after exposure to camphor vapors (WON 1950, CLARK and WEBB 1955, OGG and ZELLE 1957). In such cells the size of nuclear bodies, but not their number, is increased. These cells act as if they were "homozygous diploids" (OGG and HUMPHREY 1963, using quotation marks as shown), and survival curves in irradiation experiments also suggest di-energidy. The strains can give rise to small-cell segregant clones. The phenomenon has not been investigated at the molecular level.

H. Nucleoid Functions

1. The Regulation of Cell Division

At first sight, nucleoid division and cell division seem closely interdependent, but actually their interdependence resembles that between two cars which are proceeding in the same direction on a one-lane road. They can move and stop at will, except that the second car (representing cell division) is prevented from passing the first (representing nucleoid division). This may serve to illustrate the fact, that under normal circumstances the formation by cell division of nucleoid-less compartments is effectively prevented. (Recently a mutant F⁻-strain of *E. coli* has been isolated which is capable of pinching off nucleoid-less buds that can act as recipients in conjugation, ADLER et al. 1967.)

Delay of cell division relative to nucleoid division by one or a few cycles is a common phenomenon associated with the normal growth cycle. The fact that common bacteria during periods of rapid growth contain between two and four nucleoids per cell, has been mentioned earlier (p. 10). More specifically, the number of nucleoids per cell parallels growth rate (Fig. 2, p. 9). If rapid growth is initiated by a "metabolic upshift" (i.e. transfer to new or rich medium), RNA and protein synthesis are the first to proceed at a higher rate, followed by DNA synthesis (KJELDGAARD et al. 1958, MAALØE and HANAWALT 1961). Cell division tends to lag with respect to these events. Upon metabolic "downshift" (due to exhaustion of the growth medium or by transfer to a poorer medium), protein and DNA synthesis are slowed down or discontinued completely. In this case, cell divisions will still occur and eventually will continue until the cells have become uninuclear.

Some bacteria exhibit specific growth cycles which formally resemble the cycle described above without being fully explained by the mechanism mentioned. *Proteus* spp. forms multinuclear, non-septate filaments, "swarmers", which are formed during periods of unrestricted growth and later divide into as many cells as they contain nucleoids (KLIENEBERGER-NOBEL 1947 c, HOENIGER 1966, see Fig. 3, p. 11).

Rapid growth of *Arthrobacter crystallopoietes* in a suitable medium may or may not result in the temporary conversion of coccoid cells into rod-shaped ones during the period of most active growth. This morphogenetic effect is not only related to growth rate but also dependent on the nature of the substrate (ENSIGN and WOLFE 1964). No caryological studies have been made on this organism to show whether nutritional control affects the coordination of cell and nuclear division.

Experimentally, inhibition of cell division without concomitant inhibition of nucleoid divisions can be obtained in a variety of ways; all result in the formation of multinucleated filaments. The nucleoids may separate normally or form a centrally located nuclear body which usually is capable of separating into individual nucleoids, once normal conditions have been restored. Multinucleated filaments were obtained by treatment with urea (PULVIRENTI 1951), ultraviolet light (DEERING 1958, KANAZIR and ERRERA 1959), gamma irradiation (PONTEFRACT and THATCHER 1965, HOENIGER 1966), azaserine (MAXWELL and NICKEL 1954), d-amino acids (GRULA and GRULA 1962), by partial inhibition of protein synthesis (MARUYAMA and LARK 1961), and during magnesium deficiency (ONOFRIO and FALCONE 1957). A temporary arrest of cell division with respect to nucleoid division was also observed in early growth phases of *E. coli* in microcolonies on agar derived from old stationary cultures (PITZURRA and RAGNI 1958). ADLER and HARDIGREE (1965) obtained temperature-sensitive mutants with impaired cell division but normal nucleoid division. Cell divisions could be initiated at 42° C (not at 30° C) or by exposure to pantoyl lactone or other sulfhydryl agents (COHEN et al. 1967). Upon treatment with mitomycin, nucleoid divisions did no longer occur, and nucleoid material was accumulated in the central region of the filaments (SUZUKI et al. 1967). In the latter formations, cell divisions could no longer be induced by pantoyl lactone or by a temperature shift to 42° C.

The timing with respect to each other of nucleoid and cell divisions was affected in some way by transfer of cells into a medium containing ribosides. K. G. LARK (1960) and C. LARK and K. G. LARK (1962) obtained phased synthesis of DNA while cell divisions were not phased.

In a more general sense, cell division is affected by agents interfering with cell wall synthesis. Such effects in many cases do not directly interfere with nucleoid divisions (see p. 85).

Recent experiments by WALKER and PARDEE (1967) link septum formation (i.e. the initiation of cell division) with the operability of a certain site on the genophor. This observation points towards a cause-and-effect relationship which may be formulated in the following way: Cell division requires a specific piece of information which is generated during nucleoid replication at a rate of one piece per round. This piece is good for one cell division only, but it is stable and can be redeemed immediately or after a virtually unlimited period of time.

A long-lived message of this sort can legitimately be called a "structural element", since present usage of this term does not include any commitment as to the number and size of molecules involved.

The element under consideration does not interfere with subsequent rounds of DNA replication or nucleoid divisions. In the words of Forro (1965): "Cell division is not a necessary or sufficient condition for the initiation of a round of chromosomal replication. Neither must cell division await the start of new rounds of DNA synthesis in both daughter chromosomes. These conclusions are consistent with the well established facts of variation in size, DNA and nuclear body content shown under various environmental conditions (Schaechter et al. 1958). Thus the conjecture that regulation of cell division and chromosomal replication in *E. coli* is a mutually dependent phenomenon (Jacob and Brenner 1963) is doubtful as a generalization."—For further discussion see also Kuempel and Pardee (1963) and Roberts (1960).

Finally let us consider the role of the hypothetical element in nucleoid division. It has been pointed out earlier that the element presumably is generated during replication of the genophor, and that nucleoid division is a twofold process involving (1) the separation of the DNA daughter structures into largely independent nuclear entities. (This process invariably accompanies replication and represents nucleoid division on the molecular and sub-light-microscopic level.) This event is followed by (2) the light-microscopically visible segregation of nucleoids into separate nuclear bodies. The latter process may occur simultaneously with the former, but may be delayed until shortly before cell division. In this way, visible separation of daughter nucleoids may precede septum formation whereas in other instances the inception of septum formation may precede nuclear segregation. Obviously neither of these events is capable of controlling the other.

This does not rule out the possibility that the signal generated at the DNA acts upon some site at the membrane or cell wall. It may be that the immediate consequence of its action is the growth of the longitudinal wall, and this in turn may control both nucleoid segregation and cell division in a rather flexible manner. This idea is in agreement with the original proposal by Ryter and Jacob (1963, see also p. 30 -31).

With some modifications, the same considerations apply also to special mechanisms which lead to the inclusion of nucleoids in resting stages such as conidia and endospores.

2. Interactions between Nucleoids and Other Genetic Determinants

a) Plasmids and Episomes, General Remarks

The term p l a s m i d (Lederberg 1952) was chosen to designate a genetic determinant which is not an integral part of the bacterial genome and which acts as an "extranuclear" transmissible agent (Lederberg 1958). Jacob and Wollman (1958) introduced the term e p i s o m e for essentially the same type of accessory genetical element, the only additional requirement being that an episome may exist and replicate in the autonomous state or may become integrated into the bacterial genome and then may replicate together with and under the control of the latter. Some of the accessory genetical elements that have been discovered cannot be clearly classified as yet in terms of the plasmid-episome alternative, nor is it certain whether such classification will remain adequate as our knowledge widens. As an example of possible complications, reference shall be made to the resistance transfer factor which, according to DUBNAU and STOCKER (1964), cannot be integrated into the bacterial genome except when such integration is mediated by a prophage, P 22). For a complete survey of literature on episomes, the reader is referred to a review by CAMPBELL (1962).

b) Bacteriophage

Bacteriophage, contrary to other such elements, can exist outside the bacterial cell (i.e. as mature phage) besides being capable of replicating in the autonomous (vegetative) or integrated (prophage) state inside a bacterial cell. Integration of phage into a bacterial genome is a recombinatory event. As a consequence, the phage genome can act as a sequence of bacterial genes, conferring certain hereditary traits to the bacterial phenotype. Prophage is undetectable with present cytological techniques. The size of the DNA segment, however, is known in many instances.

Vegetative phage, arising from infection or from the induction of prophage, not only replicates at a faster rate than is characteristic of the bacterial genome, but frequently also blocks the functions of the bacterial nucleoid. Among the enzymes which are transcribed quite early in many phage developmental cycles is a deoxyribonuclease capable of degrading specifically the host DNA (KELLENBERGER 1961). If such enzyme is formed, the molecular fragments of the host DNA are utilized for the synthesis of phage DNA (STENT and MAALØE 1953, HERSHEY et al. 1954). Some differences are found among different species of phage in that some tend to destroy the bacterial genome shortly after infection while others initially do not interfere with the replication of the host genome (WHITFIELD 1962) or do not materially affect the host DNA (see below).

The fate of the bacterial nucleoid during phage infection has been studied with both the light and electron microscope. Light-microscopic observation of stained preparations revealed changes described as disintegration and subsequent increase in mass of DNA material (DELAPORTE 1950, LURIA and HUMAN 1950, HEDÉN 1951, KELLENBERGER 1961). Electron microscopy of ultra-thin sections showed that the fine structure of the DNA-plasm was essentially unaltered during nucleoid degradation and synthesis of vegetative phage. The DNA-plasm increased in mass, and its outlines became more irregular. Scattering of nuclear material throughout the cell was apparent but may not be real, since serial sectioning reveals a branched but still coherent mass of nuclear material (PARVIS 1950, MAALØE et al. 1954, WHITFIELD and MURRAY 1954, 1957, RYTER 1960, KELLENBERGER et al. 1958a, b, KELLENBERGER 1961, 1963, KRAN 1962, SHADOMY 1963). Giant DNA pools are obtained if maturation of phage is prevented by the addition of chloromycetin (KELLENBERGER et al. 1958a, b). The completeness of the incorporation of pool DNA into mature phage is proof for the degradation of the nucleoid which takes place during phage multiplication. Formation of mature phage involves condensation of DNA in a degree never observed during the normal life-cycle of any bacterium (KELLENBERGER 1963, FUHS 1964, 1965 d). The DNA inside a bacteriophage head is condensed into 1/13 of the volume which it would occupy in a fully hydrated state as found in the DNA-plasm of bacteria.

A DNA phage which is capable of multiplying autonomously without ever consuming the host genome and which upon maturation leaves the host cell without necessarily rendering it nonviable was described by HOFFMANN-BERLING et al. (1963) and by MARVIN and HOFFMANN-BERLING (1963). In these bacteria (male *E. coli*) a balanced relation is established between the bacterial and phage genomes although multiplication of the latter is favored (HOFFMANN-BERLING and MAZÉ 1964). About 50 per cent of the infected bacteria remain viable after release of mature phage but remain infected.

While a detailed account on individual phage species is beyond the scope of this review, the latter type requires consideration as a kind of virus which possibly is more closely related to the plasmagene-like determinants dealt with in the following sections.

c) Plasmids Other than Phage

Plasmids other than phage do not exist as extracellular entities. Their transmission requires cell-to-cell contact of bacteria. This process is called c o n j u g a t i o n. The elements still are capable of multiplying faster than the bacterial genome, but they never become abundant in any one cell nor does their presence in itself negatively affect bacterial viability.

The most thoroughly investigated plasmid "species" is the F episome or the *E. coli* fertility factor (HAYES 1953). Its essential constituent is a DNA segment or genophor, the size of which has been given as approximately 2.5 to 3×10^5 base pairs or 1 per cent of the bacterial genophor (DRISKELL-ZAMENHOF and ADELBERG 1963, HERMAN and FORRO 1964).

The plasmid col E_2 (a colicin) consists of a circular DNA molecule 2.32 μ long (ROTH and HELINSKI 1967).

The relationship between the bacterial nucleoid and these elements is twofold: Recombinatory events may result in the incorporation of the episomal element into the bacterial genophor. In this case the episome is replicated as a part of the bacterial genome and under the control of the latter. But also may segments of the bacterial genophor become associated with the episome. In such case the bacterial segment is replicated as well as transferred during conjugation as an integral part of the episome. Finally, in Hfr strains of *E. coli* a compound genetic structure is formed which, depending on the situation, operates under bacterial or episomal control.

The probability of integration of an accessory genetical element is at least greatly enhanced by the presence on the accessory particle of regions homologous to regions of the bacterial DNA. Whether the presence or absence of homologous regions is the only factor determining whether a particle can be integrated into the bacterial genome, is not clear. The tendency of episomes to become integrated is known to vary and appears to be a characteristic property of the episomal species. Colicinogenic factors tend to exist in the integrated form and, if integrated, do not transfer the bacterial genome. The resistance transfer factor, however, rarely undergoes recombinatory association with the bacterial genome (WATANABE 1963, PAINTER and GINOZA 1966, FALKOW et al. 1966). The *E. coli* fertility factor occupies an intermediate position.

Episomal genes determine the bacterial phenotype as do bacterial genes, regardless whether the episome is in the autonomous or in the integrated state. An episome which carries a segment of the bacterial genome renders its host diploid for this particular region of the genophor. This phenomenon has been used to investigate dominance and recessivity of bacterial genes.

Characters conferred upon bacteria by episomes are the resistance against combinations of antibiotics ("resistance transfer factor"), the production of agents causing lysis in strains not harboring the specific episome ("colicinogenic factor"), and antigenic properties. The latter have been studied particularly with reference to the episomal character of bacterial sexuality. It was found that *E. coli* mating types could be characterized immunochemically. Strains which harbor the "male" character (the F episome) and which are capable of transferring this character to "female" (F-less, F−) cells, showed a particular antigenic property (ØRSKOV and ØRSKOV 1960, SNEATH and LEDERBERG 1961). This property was considered important in promoting the establishment of cell-to-cell contact between cells belonging to opposite mating types. Subsequently it was found that this antigenic character was located on a particular type of pilus which is found exclusively on the surface of male (F+ and Hfr) strains and is also absent in F− phenocopies, i.e. cells which harbor the F episome but fail to form conjugation pairs. Acquisition of the F episome results in development of the F-specific type of piliation within less than one hour (VALENTINE 1966). These F-pili not only serve as sites of attachment of bacteriophage specific for male strains of *E. coli* (CRAWFORD and GESTELAND 1964, CARO and SCHNÖS 1966), but were also considered important in the establishment of cell-to-cell contact at conjugation and even were suspected to represent the conjugation bridge proper which serves as a duct for the F-mediated transfer of genetic material (BRINTON 1965). While this model seems to explain some phenomena of conjugation, other workers observed direct cell-to-cell contact in bacterial pairs at conjugation (WOLLMAN et al. 1956, ANDERSON et al. 1957, GROSS and CARO 1966; more pictures of "conjugation bridges" are found in the literature, also in species other than *E. coli* [e.g. BLADEN 1963], but their significance is unproven because of lack of supporting genetical data).

· In order to understand better the effect of F-factor integration into the bacterial nucleoid, we have to mention in some detail the properties of the F episome in its autonomous state. The F episome most likely is a circular DNA structure similar to but much shorter than the bacterial genophor (CAMPBELL 1962). It represents an autonomous replicative unit (replicon) in the sense of JACOB and BRENNER (1963), but its replicator system in some respect is different from that associated with the bacterial genophor.

Replication of the F element can be inhibited by addition of acridine dye in concentrations which do not interfere with the replication of the bacterial genome. An *E. coli* F⁺ strain upon treatment with acridine dye becomes F⁻ because the replication of the F factor is slowed down or inhibited completely, while the bacterial population increases at a normal rate (Hirota 1960, Stouthamer and de Haan 1963). As a result, the fraction of cells harboring the episome decreases rapidly. Continued investigations have resulted in the isolation of mutants carrying a defective F-replicator system. Among them are strains in which the F episome is not replicated at elevated temperatures while the bacterial genome is. For details the reader is referred to Jacob et al. (1963).

The mechanism regulating replication of the autonomous F factor is not clear. The observation that the F⁺-property spreads through a growing population of F⁻ cells (Lederberg et al. 1952, de Haan and Stouthamer 1963) could be explained by repeated conjugation and replication at conjugation of the F episome. This would also explain the occurrence of the F episome in numbers per cell not significantly higher than one (see Hayes 1964, p. 652 ff. for references). In agreement with this idea are the conclusions on the grounds of available evidence that frequent reinfection may be necessary for maintenance of the F⁺ state (Campbell 1962) and the observation that in certain mutants both the ability of the F factor to reproduce autonomously and its transfer activity are altered (Jacob et al. 1963). Other experiments, however, indicate that cells containing the F episome can transmit this property to progeny cells. The analysis of progeny from single F⁺ cells showed all cells to be F⁺ (Lederberg 1958). From these experiments, Hayes (1964) concludes that nucleoid and F episome may be replicated and distributed into daughter cells in a coordinated manner resembling coordination of chromosomes at mitosis (see also Jacob et al. 1963, 1966). In their 1966 paper, Jacob et al. present evidence that nucleoids and F elements segregate jointly into daughter cells. Polynucleotide strands of both elements which were synthesized in identical time intervals were found to be distributed into the same daughter cell. This is explained in terms of the attachment of both to conserved membrane segments (Fig. 11, p. 32).

Conjugational transfer of the F factor not accompanied by replication is indicated in experiments by Adelberg and Driskell (unpublished, cit. by Hayes 1964, p. 654). F⁺ bacteria that have been rendered uninuclear by growth in phosphorus-deficient medium, upon mixing with a suspension of F⁻ cells lose their episomes at the same rate as recipients are converted to F⁺.

For more details concerning episomes the reader is referred to monographs on bacterial genetics (e.g. Hayes 1964, Braun 1965).

d) Interaction of Bacterial and Episomal Replicator Systems in *E. coli* Hfr Strains

E. coli Hfr strains are derived from F⁺ strains as mutants. Genetical analysis shows that these strains contain a single DNA unit which consists of a bacterial genophor with an integrated F episome. This unit under normal conditions must be circular (WOLLMAN and JACOB 1959, JACOB and WOLLMAN 1961). This latter conclusion has been confirmed by actual observation in autoradiograms (see p. 14). But the genome of episomeless (F−) strains also is represented by a single circular linkage group (WOLLMAN and JACOB 1959). Therefore it is reasonable to assume that Hfr strains originate from strains with an autonomous F element (F+) by a single recombinatory event which results in linkage of the two circular structures (JACOB and MONOD 1963). Integration of the F factor into the bacterial genome and dominance of the bacterial replicator system during normal reproduction of Hfr bacteria are evident from the observation that such replication is insensitive against concentrations of acridine orange that will inhibit multiplication of the F particle in its autonomous state. Insertion of the F particle can occur at different sites along the genetic chart, 17 sites being identified thus far (MATNEY et al. 1964). It is not clear whether these sites are characterized by base sequences homologous to certain base sequences on the F genophor.

While DNA replication in Hfr cells normally is under the control of the bacterial replicator system, the episomal system "takes over" at conjugation. Transfer into the recipient cell of genetic material is not restricted to the episomal structure but comprises the integrated unit. Although, due to breakage of the conjugatory bridge, transfer is rarely complete, the regular appearance and integration of donor strain markers in recipient cells are characters of Hfr-F conjugation and have led to the designation Hfr (high frequency recombination) strain. Analysis of marker transfer shows that the structure transferred is linear rather than circular. The origin of replication is transferred first while the segment conferring the F+ character to the recipient cell is transferred last (if at all).

The transferred material originates from a replication process which takes place in the donor cell during transfer. One copy is transferred into the recipient cell while the other is retained by the donor (JACOB et al. 1963, BLINKOVA et al. 1965, PTASHNE 1965, HOLLOM and PRITCHARD 1965, GROSS and CARO 1965, 1966, KRISCH and KVETKAS 1966, FREIFELDER 1966, ISHIBASHI 1966). HERMAN and FORRO (1964) had found in autoradiographic experiments that the F factor transferred during conjugation was the product of prior semiconservative replication, although their experiments did not permit to decide whether replication and transfer were simultaneous processes.

From these observations it appears that DNA transfer is the result of the activity of the F replicator system. Such activity may be triggered by the establishment of contact between donor and recipient cells which may cause the formation or activation of an F-specific initiator protein. Involvement of a replicator system different from the normal bacterial system is apparent from the distinctly different speed of the replication process and its polarity, which is opposite to that during normal replication. Furthermore replication during conjugation is sensitive against acridine orange which directly points towards an involvement of the F replicator system. (Irradiation and ^{32}P incorporation can induce reversal of polarity of transfer, but it is not apparent whether the direction of normal replication is also reversed. For details see KRISCH 1965, JOSET et al. 1964.)

From the fact that establishment of cell-to-cell contact almost immediately results in transfer of the genetic material, JACOB et al. (1963) speculated that the F replicator system may be located at the cell periphery and in immediate neighborhood of the antigen which presumably is active in establishing contact with the recipient cell.

At this point, considerations on the possible organization of an *E. coli* Hfr nucleoid during conjugation reflect on models of the n o r m a l mode of replication in such cells. The idea of a genophor which is ready for transfer by permanent attachment of the F replicator to a specific membrane site is hardly compatible with the idea of permanent attachment of the same structure at the replication fork during periods of replication under the control of the bacterial replicator—a conclusion which was arrived at "by analogy" but which actually is in conflict with the former. Alternate attachment of bacterial and episomal replicator sites is a possible alternative, with both sites located on opposite polynucleotide strands as indicated by opposite directions of bacterium-controlled and F-controlled replication.

Injection of DNA during conjugation requires metabolic energy on the side of the donor bacterium; materials other than DNA are excluded from transfer (SILVER 1963). Transfer proceeds linear with time (FUERST et al. 1956, MARCOVICH 1961). The part of the recipient cell is considered a passive one, although some specific metabolic step may be involved (RAJCHERT-TRZPIL 1963).

As a consequence of conjugation, the exconjugant of an Hfr $\times$ F− cross acquires a DNA segment the size of which depends on the duration of the conjugation process. If transfer proceeds to completion, the unit acquired represents the Hfr genophor. In such case, the Hfr character is conferred to the recipient cell. In the majority of cases and depending on details of methodology (see e.g. TAYLOR and ADELBERG 1960, ALFOLDI et al. 1962), transfer is incomplete. Since apparently the F replicator system is transferred last, the recipient cell in most cases remains F−. The DNA fragment will render the exconjugant diploid with respect to markers on the corresponding region of the genophor, but since such fragment normally lacks the ability of replicating autonomously, it will be diluted out in subsequent divisions. Its markers must be transferred to the bacterial genophor by recombination in order to become inheritable characters of the bacterium (ANDERSON and MAZÉ 1957).

Exceptional strains which maintain "diploidy" in certain regions of their genetic chart were isolated by LEDERBERG (1949, 1957/58). The type of association between the parental genomes in these strains is not clear. Possible explanations are the integration of the fragment received at conjugation into a large duplicated region of the genophor, or existence of a replicator system on the segment received from the donor cell. The latter possibility was discussed by JACOB and MONOD (1963, cit. after HAYES 1964).

Little attention has been given to the state of the donor cell after conjugation, in particular with respect to the participation of one nucleoid and the exclusion of others in the process of genetic transfer. Interest in the recovery of markers in the recipient cell has been the primary matter of concern, and this usually involves elimination of donor cells after mating. The fact that the donor cells retain their F[+] or Hfr character (ANDERSON and MAZÉ 1957) was explained by replication of the genetic structure to be transferred and retention by the donor cell of one of the copies.

e) Other Systems of Conjugation

Several conjugation systems involving bacteria other than *E. coli* have been described in the literature. Some are interspecific in that sex factors from *E. coli* and related enteric bacteria can be transferred to other enterobacteria such as *Proteus* or *Serratia* or even less related forms such as *Vibrio* and *Pasteurella* (MARMUR et al. 1961 a, BARON and FALKOW 1961, FALKOW et al. 1961, 1964, MARTIN and JACOB 1962, BARON 1963) as well as Gram-positive organisms (ANTOHI et al. 1966).

A system which in some respect is similar to the *E. coli* system was found in strains of *Pseudomonas aeruginosa* (HOLLOWAY 1955, 1956, HOLLOWAY and JENNINGS 1958, HOLLOWAY and FARGIE 1960, LOUTIT and PEARCE 1965). A bacteriocin-producing transmissible factor was described in *Vibrio cholerae* by BHASKARAN (1958, 1960) and BHASKARAN and IYER (1961).

While a detailed appraisal of these systems is beyond the scope of the present review, the mechanism of genetic exchange in certain *Agrobacterium* and *Pseudomonas* species deserves consideration, since it is associated with certain cytologically conspicuous events and involves the nucleoid. As pointed out later in greater detail (p. 72), cells of these organisms in certain phases of their life cycle and under favorable cultural conditions tend to aggregate in star-like clusters. The nucleoids approach the cell ends facing the center of the star. Earlier observers correctly suspected an exchange of genetic material during star formation, although the alleged fusion of several nucleoids does not occur. The exchange of genetic material rather is restricted to single pairs and consists in the transfer of complete genomes. Polar fimbriae or pili are important in establishing cell-to-cell contact during star formation. Sexual polarity and an episomal basis of sexuality, however, have not become apparent. For details, the reader is referred to HEUMANN (1956, 1960 a, b, 1962 a, b, 1963), MARX and HEUMANN (1963) and HEUMANN and MARX (1964).

f) Other Suspected Episomal Factors and Nucleoid Participation in Other Forms of Genetic Exchange

Transformation. Transformation experiments were the first to show that DNA is the carrier of genetic information (AVERY et al. 1944, McCARTY and AVERY 1946). Since then, many workers have studied transformation in *Diplococcus, Neisseria, Bacillus* and *Haemophilus* with special emphasis on the recovery of markers from DNA preparations as well as the mechanism of DNA uptake by competent cells. Since preparations of transforming DNA consist of more or less mechanically disrupted DNA fragments, uptake of such DNA does not result in its replication, unless the fragment (or the specific marker) is integrated into the bacterial genome by a recombinatory event. Recombination occurs whether or not the recipient cell is in a state of replication (Fox 1960, BODMER 1963, see also LACKS 1962). Due to the highly statistical character of transformation, genetic studies have met with more success than efforts to study the physiological state of the bacteria transformed. Recently it has transpired that the receptive state of these cells ("c o m p e t e n c e"), at least in certain forms, besides protein synthesis may require certain nucleoid functions. JYSSUM (1965) and JYSSUM and LIE (1965) found competence in *Neisseria meningitidis* to be a strain-specific character which was apparent throughout the growth cycle but occasionally was irreversibly lost. In experiments resembling those by YOSHIKAWA and SUEOKA (see p. 17) it was found that competent and non-competent strains replicated their genophors with opposite polarities. On the basis of these observations, the existence in *N. meningitidis* of an episomal factor determining competence was proposed.

YOSHIKAWA et al. (1964) worked with two *B. subtilis* strains only one of which was transformable. They found differences in the replication pattern although the polarities of replication were identical. They concluded that in stationary cells of the transformable strain, replication is not arrested after completion of a replication cycle while it is arrested in the non-transformable strain.

To continue the enumeration of—probably superficial—resemblances between competence patterns and episomal effects, a factor may be mentioned that transmits competence among cells of transformable cocci. It is heat-labile and probably a protein (PAKULA and WALCZAK 1963, TOMASZ 1965).

DNA Excretion. Transformation involves the uptake by bacteria of free DNA from the growth medium. One may ask whether conditions exist that favor the secretion of free DNA into the bacterial growth medium. Free DNA was found in cultures of several Gram-positive and Gram-negative cocci (SMITHIES and GIBBONS 1955, CATLIN 1956, TAKAHASHI and GIBBONS 1957, CATLIN and CUNNINGHAM 1958, OTTOLENGHI and HOTCHKISS 1960). CATLIN (1960a, b) showed that such DNA may cause transformation in the strains producing it. The cultures producing DNA also produced RNA in a high-molecular form as well as a particular deoxyribonuclease (CUNNINGHAM 1959). Accumulation of the DNA occurs only under conditions unfavorable for the activity of that enzyme (e.g. abnormal salt content of the medium, lack of activating calcium). TAKAHASHI and GIBBONS (1957) and CATLIN (1960a, b) assume that the DNA originates from lysing or otherwise non-viable cells. KLEINSCHMIDT et al. (1960a) prepared ultra-thin sections of a DNA-producing *Micrococcus* and demonstrate the similarity in structure of intra- and extracellular DNA. Their illustrations suggest that DNA is actively secreted into the medium.

Sporulation. The cytology of spore formation is dealt with below (p. 73). In this context we refer to the possibility of an episomal basis of sporulation. Such basis was assumed to exist by JACOB et al. (1960) on the grounds that extremely stable asporogenous strains of *Bacillus* species were obtained that do not revert upon transformation with DNA from other non-sporogenous strains. This observation suggests that the asporogenous strains lack a common sequence of genes. SCHAEFFER et al. (1965) found sporulation to be a character transmissible by transformation with wild-type DNA and furthermore found asporogenous mutants to be different from mutants with a defective mechanism of sporulation. The latter behaved more normally in that wild-type revertants could be obtained from strains exhibiting different types of abnormalities in sporulation. This and the mutation pattern after ultraviolet exposure suggested the existence of an episomal sporulation factor. More recently, ROGOLSKY and SLEPECKY (1964) obtained a 33 per cent yield of asporogenous mutants in *Bacillus subtilis* during early growth in the presence of acriflavin. This was interpreted as indicating the elimination of an episomal factor which normally during exponential growth becomes integrated into the bacterial genome. The authors point out that an alternative explanation cannot be excluded in view of the mutagenic action of the acridine dye.

3. Transcription

a) General

While a detailed account of current concepts of gene action and its regulation is beyond the scope of this review, the ability of the nucleoid to act simultaneously as a replicating and a working nucleus deserves consideration. Certain phenomena common to both replication and transcription have been mentioned earlier (p. 20).

The nucleoid has been established as the site of RNA synthesis with both autoradiographic and biochemical techniques (CARO 1961, FRANKLIN and GRANBOULAN 1963, GODSON and BUTLER 1962, 1963, 1964, BARR and BUTLER 1963), although some authors report that newly formed RNA appears with some preference in fractions containing membrane material (SUIT 1963, SPIEGELMAN 1959, ABRAMS et al. 1964). NISMAN and FUKUHARA (1960) found stimulation of the enzyme β-galactosidase in cell-free preparations in the presence of DNA.

Transcription of the genetic message into messenger RNA involves one strand only of the native DNA. Evidence for this was found in phage systems (HALL et al. 1963, MARMUR and GREENSPAN 1963, MARMUR et al. 1963, HAYASHI et al. 1963 a, b, 1964, TOCCHINI-VALENTINI et al. 1963) as well as in bacteria. In the latter systems only one half of the DNA (i.e. one strand of the DNA) was found to anneal with m-RNA synthesized under its control (AGENO et al. 1965). M-RNA is synthesized by addition of nucleotides at the 3' end of the chain (GOLDSTEIN et al. 1965). This finding might help to identify the template strand of the DNA if the direction of transcription is known.

Current concepts imply a basic resemblance of transcription and replication processes in that both require localized loosening of the helix struc-

ture, presumably involving single-strand breaks (see p. 21). These conclusions are also derived from the discovery of a DNA-RNA complex which
appears during transcription and which is of transient nature. The complex
resembles two-stranded DNA in that it is susceptible to denaturation by
heat or by formamide and in its insensitivity against ribonuclease (Hayashi
1965, Hayashi and Hayashi 1966, Armstrong and Boezi 1965. see also Spiegelman et al. 1961).

Electron micrographs indicate the existence of a temporary association
of DNA and RNA and also suggest a transient linkage of DNA and ribosomes by a strand of informational RNA (Bremer and Konrad 1964. Byrne
et al 1964, Bladen et al. 1965). This model implies that m-RNA synthesis
and its transcription into a polypeptide sequence proceed with identical
polarities. This was found to be correct (Smith et al. 1966, Guest and
Yanofsky 1966).

b) The Action of Chloromycetin

Chloromycetin (chloramphenicol) in appropriate concentrations markedly affects various steps in the sequence described in the preceding section
and causes changes in nucleoid chemistry and structure.

Chloromycetin interferes with the ribosome-mediated synthesis of protein and in particular with the formation of a ribosome subunit consisting
of protein. In the presence of the inhibitor, incomplete ribosomes (chloromycetin particles) are formed which contain ribonucleic acid and a protein
fraction, the synthesis of which apparently is not affected by the antibiotic
(Aronson and Spiegelman 1961, Kurland et al. 1962, Shakulov et al. 1963,
Nomura and Hosokawa 1965). Chloromycetin does not interfere with the
synthesis of messenger RNA but it interferes with its ultimate destruction,
probably by impairing ribosome functions. Therefore m-RNA tends
to accumulate in chloromycetin-inhibited cells (Hahn and Wolfe 1962,
Levinthal et al. 1963, Armstrong and Boezi 1965).

Chloramphenicol increased markedly the RNA content of nuclear bodies
isolated with the lipase technique of Spiegelman et al. (1958) (Ezekiel 1961,
1964), but the increase in RNA content failed to become apparent in electron micrographs of the nuclear bodies. While the nature of the RNA was
not determined in these experiments, Lark (1963) speculates that, under
the influence of the antibiotic, the complementary RNA may remain
attached to the DNA and in this manner may ultimately interfere with
DNA replication. This may explain, why in the presence of chloromycetin
a limited amount of DNA is synthesized before inhibition occurs (Goodgal
and Melechen 1960, Doudney 1961 a). For more references concerning this
point see p. 25.

Fuhs (unpublished results) found that chloromycetin markedly affected
nuclear structure in sectioned bacteria. *Bac. subtilis* was grown in aerated
1 per cent tryptone water with 0.5 per cent sodium chloride added. When

Fig. 14. *Bacillus subtilis* treated with chloromycetin as described in text. Fixed in osmium-dichromate mixture,
treated with uranyl acetate, embedded in Vestopal. Serial sections of a nuclear body presumably representing one
nucleoid. See text for interpretation. Scale marker represents 0.5 μ.

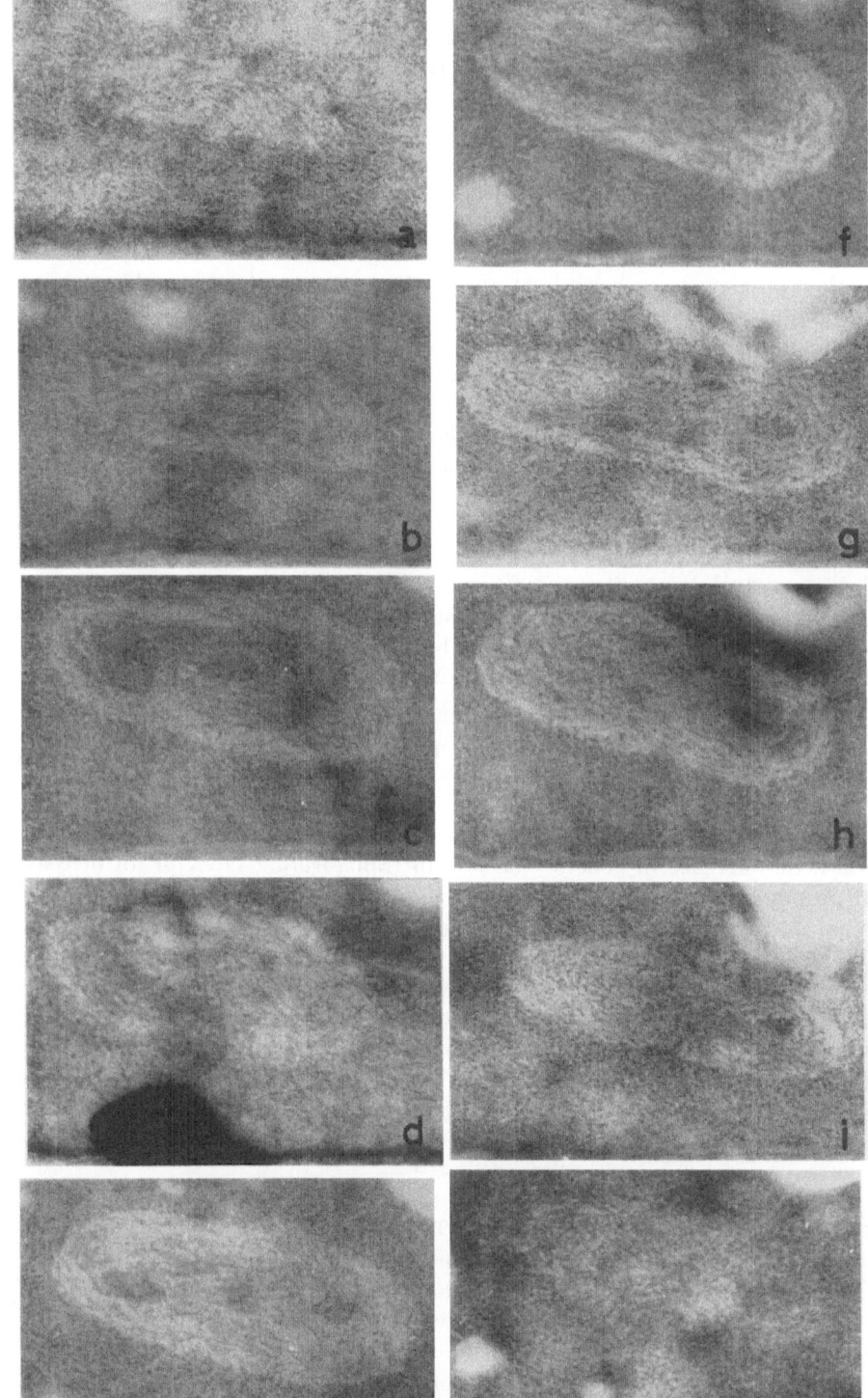

the cell count had reached 3×10^7 per ml, chloromycetin was added to give a final concentration of 20 μg/ml. After 135 minutes incubation at 30° C, DNA content had increased by 66 per cent. If such cells were fixed with osmium-dichromate, treated with uranyl ions, embedded in Vestopal and sectioned, a nuclear fine structure could be observed (Fig. 14) which was markedly different from that in control preparations (see e.g. Fig. 36, p. 93). Compared with the control preparation, which was derived from a slowly growing or stationary culture, the fibrous DNA appears less distinct or almost feather-edged due to the presence of less rigidly ordered fibrous material which also stains heavily with uranyl ions and therefore presumably is RNA. The less distinct appearance of the DNA is similar to that found during rapid exponential growth (Fig. 50, p.109). No accumulation of uranophilic material around the nucleoid as described by MORGAN et al. (1967) was observed in our chloromycetin-treated bacteria. The abnormal appearance of the cytoplasm may be explained in terms of impaired ribosome structure.

The round shape of the nuclear bodies is another effect of chloromycetin treatment. The whorled interior structure of the nuclear areas shows that the round bodies are derived from more extended formations. Rounding off and even fusion of nuclear bodies under the action of the drug were also observed in the light microscope including phase-contrast observations of living bacteria (BERGERSEN 1953b, HAHN et al. 1957, SCHAECHTER and LAING 1961).

c) Replication of DNA and Transcription in the Light of Gene-Dosage Experiments

Earlier in this review (p. 17) we mentioned gene-dosage experiments in which the increase in genetic markers in transforming DNA or an increase in prophage was used to explore the sequential mode of replication of the bacterial genome. In these experiments, the indicator system operated on the genetical level. Modifications in quantitative respect of gene activity by regulatory mechanisms were excluded from considerations. Other workers synchronized cell division in bacterial cultures and found a certain periodicity of gene functions or of enzyme synthesis. This prompted speculations as to whether transcription was a sequential process similar to replication and also whether both processes might be coordinated, in that sections of the genophor undergoing replication are barred from acting as templates for RNA synthesis.

Contrary to earlier findings on the periodicity in the synthesis of β-galactosidase (BARNER and COHEN 1956, MARUYAMA 1963), CUMMINGS (1965) found the synthesis of that enzyme in its fully induced state to be as continuous as the synthesis of the nucleic acids, provided that a technique of synchronization was employed that reduced to a minimum any stress imposed on the cells. These results indicate that the site on the genetic map held by β-galactosidase is available for transcription during most if not all of the replication cycle. This, however, holds only if the enzyme is fully induced. Transcription of a repressed gene requires DNA replication, and replication of a repressible gene invariably is accompanied by its transcription also in the absence of inducer (McCarthy and Bolton 1964, Hanawalt and Wax 1964). One may speculate that the unavailability of repressor for one of the copies causes it to be transcribed just once, until a newly formed repressor molecule closes the site for repeated transcription.

Recent gene-dosage experiments based on the determination of gene activity have shown that the c a p a c i t y of a bacterium to synthesize a specific enzyme doubles at the very moment the replication fork passes along that specific site on the genetic map (Masters et al. 1964, Masters and Pardee 1965, Kuempel et al. 1965). The term "capacity" stands for the m a x i m u m p o s s i b l e rate of synthesis as obtained under suitable conditions. Changes in the a c t u a l rate of synthesis as determined by the repression mechanism can occur at any time during the cell cycle, and the observed cyclic appearance of certain enzymes is not immediately related to the replication cycle of the DNA. Similarly, the sequence of messenger ribonucleic acids generated during spore germination is not correlated with the process of DNA replication (Halvorson 1967). For the explanation of cyclic changes in enzyme level see Donachie (1965) and Masters and Donachie (1966).

A somewhat puzzling phenomenon is the production of ribosomal RNA which is believed to be copied directly from the DNA. The genetic sites for ribosomal RNA have been investigated by Dubnau et al. (1965). It was observed that this type of RNA was synthesized during certain periods of the cell replication cycle. In Hfr strains of *E. coli* which replicate their genomes with opposite polarities, the pulses of ribosomal RNA produced under the control of two distinct genetic sites appeared in the reverse order. This was taken as an indication that transcription occurred in the same order as replication (Rudner et al. 1964, 1965). The synthesis of ribosomal RNA appears to resemble the synthesis of repressed enzymes as discussed above. This is of interest with respect to the dynamics of ribosome turnover, since ribosome content of bacteria parallels growth rate (Ecker and Schaechter 1963 a, b).

In summing up: The bacterial nucleoid is capable of operating simultaneously as a replicating and as a "working" nucleus. This is in accordance with the uniformity of its chemical and organizational state throughout the cell cycle. Furthermore, mechanisms controlling transcription are not affected by the replication process.

This is plausible also on the following grounds: The bacterial genophor exhibits a twofold polarity. The first kind of polarity is that of transcription, i.e. the polarity of the cistrons and of the synthesis of messenger RNA. It appears that messenger RNA is copied from one of the strands only and that the function of the complementary strand is that of a stabilizer and of a template during replication of the duplex (Hayashi et al. 1963 a, 1964). From this and from the intricateness of the genetic message one may conclude that the template strand and the polarity of transcription are quite rigidly determined on the cistron (or base sequence) level. It is not essential from this point of view that template strand and polarity of transcription be identical over an extended region of the genophor although such may be the case. Contrary to transcription, the replication process by definition proceeds along the entire genophor. It does not involve recognition of the genetic message proper. Both starting point and polarity of replication are determined by the site of attachment or activation of the replicator element. Obviously, transcription appears to "trail" replication under certain circumstances, but it does not necessarily, on the cistron level, proceed with a polarity identical with that of replication.

J. The Chemical State of the Bacterial Nucleoid

The chemical state of bacterial DNA *in vivo* has received relatively little attention, although its investigation undoubtedly would reveal important properties of the nucleoid. Recent speculations on the role of histones in eucaryotic cells (see e.g. Barr and Butler 1963) lucidly illustrate possible implications of nucleoid chemistry.

Phase-contrast and electron microscopic observations show that the dry mass content of bacterial nuclear bodies is lower than that of the surrounding cytoplasm (Winkler and Knoch 1951, for more references see p. 56, 87). This is due to the highly hydrated state of the bacterial DNA-plasm which in turn is caused by the highly hydrophilic character of the DNA. Responsible for this property are mostly the phosphate groups which are located at the periphery of the DNA helix and which carry one negative charge each. In the acid form, DNA is highly dissociated. Hammarsten (1924), assuming a tetranucleotide structure, gave 4.3×10^{-3} as the first constant of dissociation. In fact, native DNA in its acid form is unstable due to the repulsive forces of the free phosphate groups and undergoes strand separation ("acid denaturation"). This means that the stability of the helix with respect to strand separation is greatly affected by salt and complex formation involving the phosphate backbone. As an alkali salt, native DNA is stable, although in very dilute solutions dissociation of the ligands from the polyanion is favored and will result in denaturation (Felsenfeld and Huang 1960). Particularly weak is the complex with potassium; the sodium and lithium complexes are slightly stronger (Ross and Scruggs 1964). The basic amino acids, arginine and lysine, are not stronger bound than is sodium (Felsenfeld and Huang 1960).

Much stronger than all these are complexes with divalent metals such as magnesium, calcium and manganese, and with di- and polyamines of the type $NH_2(CH_2)_nNH_2$ and of the spermine and spermidine type, as well as complexes with the polyamino acid, polylysine (Felsenfeld and Huang 1959, 1961, Jordan 1951). The association constants of these complexes are two or three orders of magnitude higher than those of the alkali metal complexes. Exact figures are not easy to provide, since many authors failed to give due consideration to nearest neighbor effects in the neutralization of the polyanion, and because binding is dependent on the ionic strength of the medium. An association constant in the range of 10^3 to 10^5, however, can be assumed to apply for the binding of the first ligand ion to the "empty" polyanion at ionic strengths around 0.1. For details the reader is referred to WIBERG and NEUMANN (1957), ZUBAY and DOTY (1958), ZUBAY (1959), SHACK and BYNUM (1959), DEKKER (1960), FELSENFELD and HUANG (1960), HUANG and FELSENFELD (1960), MAHLER et al. (1960), FRASER and MAHLER (1962), TABOR (1961), BANERJEE and PERKINS (1962). With diamines, a chain length of $n = 5$ is optimal for the binding to native DNA; the distance of the amino groups in this case equals the distance of the phosphate groups (MAHLER 1963, MEHROTRA 1964). LIQUORI et al. (1967) developed a model of the DNA-polyamine complex.

A third group of complexes, not occurring *in vivo*, is formed with the strongly bound ions of uranium (UO_2^{++}, uranyl) and lanthanum (La^{+++}). The association constants of these complexes are in the order 10^5 to 10^6 (ZOBEL and BEER 1961, BEER and ZOBEL 1961, WOLFE et al. 1962, FUHS 1965 c, STERN and STEINBERG 1953). In RNA, lanthanum and some other metals such as zinc and lead, can cause cleavage of the polynucleotide chain, but this activity is very slow or undetectable with DNA (BAMANN and TRAPMANN 1959, WEYGAND et al. 1951, BRITTEN 1962, BUTZOW and EICHHORN 1965, EICHHORN and BUTZOW 1965).

The ions of mercury, silver and copper preferentially react with the amino groups inside the DNA helix and tend to loosen the helical structure (YAMANE and DAVIDSON 1961, 1962, EICHHORN 1962, KATZ 1963, VENNER and ZIMMER 1966). The destabilizing action of divalent copper is particularly pronounced (WEED 1963, ROPARS and VIOVY 1964, VENNER and ZIMMER 1964, EICHHORN and CLARK 1965, HIAI 1965, FUHS 1965 b).

These relationships are of great importance in bacteria, since bacterial DNA does not normally occur in an organized association with protein, but must be assumed to be more or less completely neutralized by some of the small molecules and ions mentioned above. The absence of deoxyribonucleohistone is characteristic for bacteria and Cyanophyceae and apparently is shared in the eucaryotic group of organisms only by dinoflagellate (RIS 1962, DODGE 1964 b). (According to these authors, the dinoflagellates are not to be considered typical representatives of the eucaryotes. DODGE introduces the term "mesocaryotic" to characterize the nuclear status of these forms.) The difference in the chemical state of DNA in proto- and "meso"-caryotic cells as compared with eucaryotic cells is easily seen in ultrathin sections. The pattern of 35 to 40 Å thick fibres of nucleohistone and of 100 Å wide double fibres which is characteristic of eucaryotic nuclei (RIS 1956, MOSES 1956, PAPPAS 1956, CHARDARD 1960), has never been observed in protocaryotic cells. A pattern of individual DNA molecules in a more or less aggregated form appears instead (see RIS 1961, RIS and CHANDLER 1963, KELLENBERGER 1963). RIS (1962) removed the histone moiety from a preparation of deoxyribonucleohistone and observed that the preparation changed its appearance and began to resemble the DNA plasm of bacteria, Cyanophyceae, or dinoflagellates. The typical X-ray diffraction pattern of nucleohistone could not be detected with preparations obtained from bacteria with methods normally used for the isolation of deoxyribonucleo-histone (WILKINS and ZUBAY 1959 a, b, ZUBAY and WATSON 1959).

Several investigations of isolates of bacterial DNA containing protein showed that the protein fraction was not of the histone type (MASUI et al. 1962, TSUMITA and CHARGAFF 1958, KADOYA et al. 1964, BUTLER and GODSON 1963). The latter authors in particular provide reliable evidence, since ribosomal material was recovered almost quantitatively in a different fraction. Ribosomes, upon degradation, may yield materials which show some affinity towards nucleic acids in general, including DNA.

Several authors contend that the occurrence of protein in their preparations of bacterial DNA is more than accidental (PALMADE and VENDRELY 1956, PALMADE 1961, BAGHAVAN and ATCHLEY 1965), but reports on the reproducibility of these techniques and the molecular organization of said complexes are lacking, and the suspicion is difficult to dispel that these associations may be biochemical artifacts. Recently a histone-like protein fraction was found in *Staphylococcus epidermidis*. This was associated with a lipid component, presumably phosphatidic acid (BERGH et al. 1965). This result indicates that basic proteins in bacteria may occur in association with fractions other than nucleic acids. A different fraction of basic protein was isolated from *E. coli* (HURST and TAYLOR 1965).

This situation leaves little choice but to assume that bacterial DNA is neutralized *in vivo* by ions of low molecular weight, and that the exchange of ligands may proceed rather freely as determined by the availability of suitable ligands and the laws of mass action. Most probable candidates for DNA complex formation are calcium, magnesium and, in some groups of bacteria, polyamines.

Prior to their discovery in bacteria, polyamines were detected in bacteriophage (AMES et al. 1958, AMES and DUBIN 1960). Their association with DNA is apparent from the fact that they are injected into the host bacterium along with the DNA. In certain phage particles with a sufficiently permeable head membrane, polyamines can be exchanged against magnesium ions. Phage T 4 was found to contain putrescine (1,4-diaminobutane), spermidine, and magnesium ions; a *Salmonella*-phage, PTL-22, contains spermine. In T 4, polyamines may neutralize one third to one half of the DNA phosphate groups.

In bacteria, the distribution of polyamines varies. They are present in many Gram-negative organisms sometimes in considerable quantities. In Gram-positive bacteria, they are absent or preferentially located in the membrane (HERBST et al. 1958, BACHRACH and COHEN 1961, TABOR et al. 1961, TOSCHI 1963/64). Putrescine and spermidine are the most commonly encountered bacterial polyamines; monoacetyl diaminobutane and acetyl derivatives of spermidine were also found. If spermine is provided externally, intracellular accumulation of this material and its derivatives will result. KIM (1966) found that putrescine was the only polyamine harbored by a *Pseudomonas* strain. In cell-free extracts, 80 to 90 per cent were found in the soluble fraction, the remainder was present in the ribosomal fraction. This indicates that putrescine was primarily associated with the DNA and with non-ribosomal RNA. The intracellular concentration of putrescine, which under normal conditions was 0.01 molar, was reduced by 50 per cent by nitrogen starvation. In another strain, putrescine could be eliminated completely under nitrogen limitation, provided that spermidine was offered in exchange.

It can be expected that the relative mobility within the cell of these compounds renders any evaluation of their role extremely difficult, since any biochemical attack can result in their dislocation. FUHS (1965 c) therefore used an indirect approach based on the affinity of bacterial DNA towards uranyl ions. As pointed out below in greater detail (p. 89), collapse of nucleoid structure during dehydration can be prevented by half-saturating DNA phosphate groups with uranyl ions after fixation and prior to dehydration. In studying the reversibility of uranyl complex formation, the author found that after a first application and subsequent removal with disodium ethylenediamine tetraacetate of uranyl ions, the affinity of the DNA towards uranyl had increased significantly in that the concentrations of uranyl salt required for stabilization of nucleoid structure were one to two orders of magnitude lower than they were during the first application. Pretreatment with ethylenediamine tetraacetate alone produced a slighter but also significant increase of nucleoid affinity to uranyl ions. It was thought that the first uranyl treatment may displace certain ligands from the DNA phosphate groups which at this point are not removed from the cell. This latter fact was assumed to account for the original low affinity of the DNA (by competition between the existing and the introduced ligand species). Chelator treatment then would remove both uranyl ions and the original ligands. A second application of uranyl salt would find the readily dissociating sodium salt of DNA, and stabilization could be achieved with very low concentrations of uranyl ions. Rapid washout of the ligands from osmium-fixed cells implies that they consist of small molecules. The partial increase in affinity caused by treatment with disodium ethylenediamine tetraacetate alone indicates that the ligands are not exclusively metallic, since divalent metals would be removed completely by the chelator. Organic bases such as polyamines thus may be involved. This concept is supported by two observations: While during the first application of uranyl ions no nitrogenous material was removed from the cells, such material was released in great quantities during subsequent treatment with the chelating agent. Secondly, if after the presumed ligand removal by uranyl and chelator treatments the cells were exposed to a mixture of divalent metals and polyamines, the affinity of the DNA towards uranyl was decreased to almost its original low value. This shows that these ions are possible candidates for the neutralization of bacterial DNA *in vivo*. Moreover, a determination of the association constant of the *in vitro* uranyl complex and approximate quantitative considerations suggested that the association constant for the assumed *in vivo* complex is about one order of magnitude lower. This points towards the participation of calcium, magnesium, and polyamines in the *in vivo* complex.

In earlier observations, FUHS (1964) found the stabilization of nucleoids in fixed cells of *B. subtilis* to be an all-or-none event. Among cells of a random population, however, uranyl ion activities required to induce stabilization varied over a one to four ratio. This ratio was reduced after ligand removal as described in the preceding paragraph which seems to indicate that random or periodical fluctuations in the chemical state of DNA may occur.

The chemical state of DNA is an important datum in a variety of respects. The physical properties of the DNA-plasm will strongly depend on the association of the DNA with counter-ions. DNA concentration in the nucleoid is close to 1.5 per cent (weight/volume). If the DNA were present as the sodium salt, the nucleoid would be a solid gel with overlapping solvation shells of neighboring DNA segments. Association with ligands that are stronger bound than sodium will reduce the width of the solvation shells and render the segments of the genophor mobile with respect to each other which will facilitate movements and changes in shape associated with rapid DNA replication and segregation of genomes.

KELLENBERGER (1961, 1962) refers to the chemical state of DNA as a potentially important mechanism for the condensation of phage DNA into phage heads. This idea is certainly suggestive although no mechanism has been discovered thus far that resembles the association DNA with protamine during spermatogenesis in eucaryotic cells.

The chemical state of intracellular bacterial DNA apparently is of little importance for the stabilization of its native two-stranded structure. Displacement of ligands by uranyl and ethylenediamine tetraacetate treatments does not render the DNA of fixed cells more susceptible to denaturation (*Bac. subtilis, E. coli,* FUHS 1965 a). Neither is the resistance of bacterial DNA against thermal denaturation a function of its base content (see MARMUR 1960, CRAVERI et al. 1965), but is due to the absence of free ends of polynucleotide strands along the contour of the genophor and to the ability of the living bacterium to repair single strand breaks which occur during normal replication and transcription or may occur as thermal breaks at elevated temperatures (EIGNER et al. 1961).

II. Nucleoid Morphology with Emphasis on Light Microscopic Aspects

A. Introduction

The light microscope as a tool of investigation has governed fifty out of seventy years of bacterial caryology. Its limitations with respect to the analysis of nucleoid fine structure and organization have become apparent in recent years, when the electron microscope has become a valuable tool in bacterial fine structure investigation. On the other hand, certain aspects of nucleoid morphology are difficult to deal with in the electron microscope, since extensive use of the serial sectioning technique is required. In this respect, light microscopy still can provide information not easily obtained with the electron microscope, and the variety of bacterial species and physiological states of bacteria investigated with the light microscope exceeds those analysed with the electron microscope. Furthermore, light microscopy has been developed to a point where preparational artifacts can be avoided, since the mechanisms of preparation are quite well understood. This aspect still is dealt with quite lightly in many areas of electron microscopy.

In the following sections an attempt is made to present a critical reappraisal of light microscopic work on nucleoids with particular emphasis on the reliability of methods and the validity of observations if compared with available electron-microscopic and molecular evidence.

B. Historical Aspects

Light-microscopic observations date back as far as 1863 and 1866 when SURINGAR described nucleus-like bodies in *Sarcina ventriculi.* ZETTNOW (1899) found nucleus-like structures in bacteria with a technique now considered one of the most useful in bacterial caryology, i.e., a modification of the Romanowski blood stain. ZETTNOW also reports on difficulties in differentiating nucleoids and polyphosphate granules, a discrimination which has not always been made by later workers. Nucleoids were also successfully demonstrated by VEJDOVSKÝ (1900, 1904) in paraffin sections of a crustacean (*Gammarus*) that harbors symbiotic bacteria. His and ZETTNOW's papers reflect the often violent controversies that characterized the field before reliable cytochemical criteria became known and specific techniques became available. JOHNSON (1912) and RŮŽIČKA (1913) attempted to stain nucleoids in bacterial spores, recognizing that persistence of nucleoids throughout the developmental cycle was an important criterion of their nuclear character. Successful in demonstrating nucleoids in various other bacteria were also STOUGHTON (1930), THOMAS (1932), and PETTER (1933 a). Early results were summarized by MEYER (1912), DELAPORTE (1939/40), and ROBINOW (1956).

Landmarks in bacterial caryology are characterized by the introduction of improved methods of observation. Particularly stimulating was the consistently successful application of the Feulgen nuclear reaction in smaller bacteria since STILLE (1937) and PIEKARSKI (1937) and of a modification thereof, the HCl-Giemsa method (PIEKARSKI 1937, 1939, ROBINOW 1942), the first application of nucleases for the differentiation of ribo- and deoxyribonucleic-containing cell constituents (BOIVIN et al. 1947 a), the revival of direct staining methods based on modifications of the Romanowski method (BADIAN 1930, 1933, PIEKARSKI 1937, PIÉCHAUD 1949, 1951), and the improvement of phase-contrast microscopy by the introduction of compatible immersion media (BARER and Ross 1952). Details of these methods will be given in the following section.

Indirectly, the study of nucleoids was aided considerably by the development of specific histochemical techniques for the identification of cell inclusions other than nucleoids, e.g. techniques for the demonstration of polyphosphate inclusions, of cell septa, and lipid inclusions.

C. Methods of Light-Microscopic Observation

1. Observation of Living Bacteria

As a rule, nucleoids are not recognized in unstained bacteria in the bright-field microscope. The absorption of visible light in the nuclear bodies is quite similar to that in the cytoplasm. Ultraviolet light is strongly absorbed by the nuclear bodies due to their DNA content, but the same holds for the RNA-containing cytoplasm.

In cells loaded with RNA the cytoplasm may absorb ultraviolet light more strongly than the nucleoids (see e.g. HEDÉN 1951). In cells which are low in cytoplasmic RNA a positive image of the nuclear bodies is obtained (KRUIS 1913, PIEKARSKI 1938, 1939, MALMGREN and HEDÉN 1947). Ultraviolet microscopy, once considered promising also because of the higher resolution attainable, is now outperformed by refined techniques of light and electron microscopy.

While nucleoids are quite inconspicuous as amplitude objects, they are excellent phase objects. As a result of the lower solids content of the DNA-plasm the refractive index of the nuclear bodies is lower than that of the cytoplasm. If we attempt to quantitate such statement, the situation appears approximately as follows: If we assume the increment of refraction for one gram of protein or nucleic acid per 100 ml to be 0.0018 and the dry content of a bacterium to equal 30 per cent (BARER and JOSEPH 1958), the refractive index of the bacterial cytoplasm would be approximately 1.39. The nuclear bodies contain approximately 1.3 per cent (w/v) of DNA. (The DNA is calculated as sodium salt; the presence of divalent metals or poly-anions would not significantly alter this figure.) The refractive index of the nuclear material thus would be approximately 1.34.—From this, the optical path difference caused by a nuclear body of 0.4 μ thickness is obtained as $(1.39 - 1.34) \times 0.4\,\mu = 0.02\,\mu$ or 20 nm or $0.036 \times \lambda$ (for $\lambda = 550$ nm). Since the phase-contrast microscope is capable of transforming optical path differences up to $\frac{1}{4}\,\lambda$ into differences in brightness in a rather straightforward manner, it should be relatively easy to find the nuclear sites of bacteria in the phase-contrast microscope. In positive phase contrast they should stand out as brighter, in negative phase contrast as darker areas.

This was indeed the type of result obtained by earlier workers who found a quite satisfactory correspondence between the brighter areas (in positive phase contrast) and the nuclear sites as demonstrated with the Feulgen reaction (KNÖLL 1944, KNÖLL and ZAPF 1952, 1954, TULASNE 1949 a, b, STEMPEN 1950).

An important part of any phase-microscopic investigation is the choice of a proper immersion medium. Embedding bacteria in water or low solids content agar media does not produce a phase-contrast image of optimum quality, since the difference in refractive index of the bacterial cytoplasm and the embedding medium produces fringes of diffraction at the circumference of the cells which interfere considerably with the recognition of intracellular structures. This effect is eliminated by the use of embedding media the refractive index of which resembles closely that of the bacterial cytoplasm. In addition, these media must be non-toxic, non-permeable, of low osmotic pressure, and of high optical purity. Thus far only protein solutions have been found to fulfill these requirements. Gelatine was used in concentrations up to 30 per cent; equally suitable and available in even higher concentrations are solutions of bovine serum albumin (fraction V). Acacia gum was also tried but was found unsuitable for work with bacteria. The basis for this decisive improvement in the phase-contrast microscopy of bacteria was provided by the work of Barer and Ross (1952), Barer

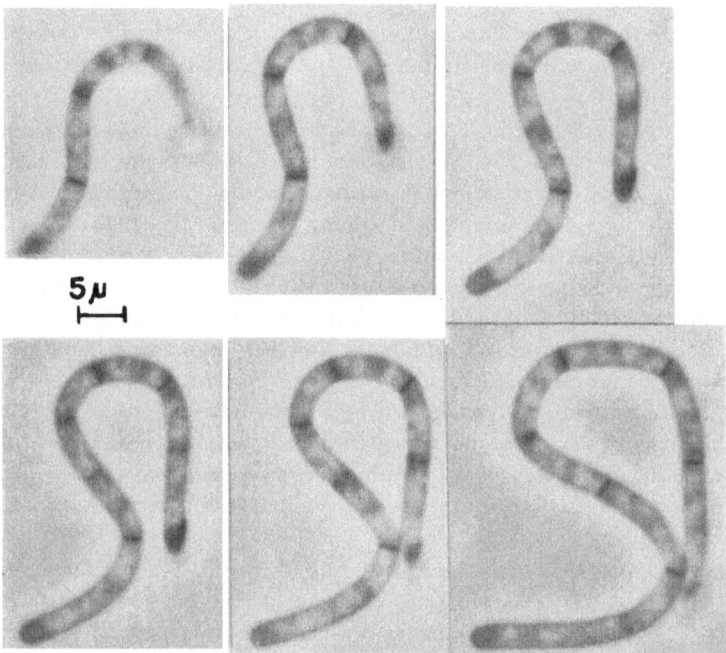

Fig. 15. Sequence of phase-contrast micrographs of growing chain of cells of *Bacillus sp.* Nucleoids show a variety of patterns which are part of the division cycle. — Courtesy Dr. C. F. Robinow.

(1952), and Barer and Joseph (1955). The results obtained by Mason and Powelson (1956), Robinow (1960), Fikhman (1963) and Adler and Hardigree (1965) amply document the value of the improved technique (Fig. 15).

No exact quantitation of phase differences has been attempted so far, since interference microscopic equipment which is necessary for such investigations has hardly ever become available to biologists. This situation is most likely to change in the very near future.

2. Demonstration of Nuclear Bodies in Stained Preparations

a) Specificity and Interferences

The difficulty of staining bacterial nuclear bodies is only partly due to their small size. The greater difficulty by far is the presence of interfering basophilic cell components.

One of these is ribonucleic acid (RNA). The RNA content of bacteria is immediately related to their rate of protein synthesis and growth as determined by the growth medium. Under conditions of slow growth, especially during nitrogen-limited growth, cytoplasmic basophilia (caused by RNA) is reduced (KNAYSI and BAKER 1947, DUGUID 1948, WILKINSON and DUGUID 1960). By the same token, agar plate cultures in which growth tends to be slower after a short initial burst have been quite popular at all times among bacterial cytologists.

Unfortunately, the advantage of nitrogen-limited growth for the successful demonstration of nuclear bodies is more than compensated by the tendency of the bacteria to accumulate in these conditions polyphosphates and lipids. Both cell constituents may interfere with the demonstration of nuclear material.

Granular polyphosphates beyond doubt are the most obnoxious of all constituents interfering with the demonstration of nuclear material and can boast a long history of confusion with nuclear elements. Polyphosphate granules (volutin, metachromatin, polar bodies) are granules of variable size which either are solid concretions of potassium polyphosphate (linear type) or are cytochemically indistinguishable from the latter. The linear array of negatively charged phosphate groups renders the granule basophilic and causes its metachromatic staining with certain basic dyes. These properties, although in somewhat lesser degree, are shared by the phosphate backbone structure of the nucleic acids, which accounts for the similar histochemical behavior of both types of compounds. Polyphosphates, however, are the more strongly basophilic compounds and retain a basic dye even at a pH value around 1, while nucleic acids are decolorized at pH values around 3.5.

KNAYSI (1942, 1955 a, b) attempted to differentiate nuclear bodies from other cell constituents by staining with methylene blue solutions of carefully adjusted pH (around 3.5). His results show that false positive results were caused in the presence of polyphosphate granules. In fact, only tests for the most basophilic of the compounds in question (i.e. polyphosphate) can be designed on this basis. Examples are the reactions for polar bodies (polyphosphate granules) with acidified basic dye solutions or the volutin test by MEYER (1904) which consists in staining with methylene blue and subsequent differentiation with 1 per cent sulfuric acid solution. Another test specific for high-polymer phosphate inclusions uses their ability to form an insoluble lead salt at pH 1 (EBEL and COLAS 1954). The fixed bacteria are treated with a 10 per cent solution of lead nitrate in N/10 nitric acid and, after extended rinsings in tap water, conversion of the deposit into the black compound, lead sulfide, with the aid of ammonium sulfide solution.

Other basophilic sites that may interfere with the demonstration of nuclear bodies are cytoplasmic sites adjacent to the membrane (plasmalemmosomes?) (BISSET 1952 a- d, 1953 b, 1954 a, b, BERGERSEN 1953 a). BRADFIELD (1956) believes that membrane-bound RNA might be involved in this phenomenon. To avoid misinterpretations, it is recommended that nuclear preparations be mounted in water which renders the cell contour clearly visible, or that membrane-stained preparations be used as control specimens. For suitable methods see PRINGSHEIM and ROBINOW (1946), BISSET (1954 a), BISSET and HALE (1953), HALE (1958), CASSEL (1951 a) and KANEMASA 1962). The polar cytoplasmic regions of some rod-shaped bacteria exhibit some tendency to stain more deeply with basic dyes than does the remainder of the cell. These areas must not be confused with nucleoids which usually are located at some distance from the cell poles (for details concerning bipolar staining see CASSEL and HUTCHINSON 1955). BISSET (1952 a) points out that shrunken protoplasts have been mistaken for basophilic cell inclusions, particularly in *Mycobacterium tuberculosis*. Such artifacts are frequent in "diagnostic" specimens which are permitted to dry out during processing.

An important point to be considered in caryological as well as genetic studies of bacteria is the multicellularity of many species and certain phases (R-forms, BISSET 1948 b). Disregard of multicellularity among other factors has contributed to the conception of the erroneous idea of mitosis in bacteria (DeLAMATER and MUDD 1951, DeLAMATER 1962, see BISSET for criticism).

But also accepted and useful techniques including such considered specific for DNA have their difficulties and pitfalls as will be pointed out below. Therefore it is imperative that any worker in the field familiarize himself with more than one method for the demonstration of nucleoids and use also methods that are specific for cell constituents that may interfere with the demonstration of nuclear material.

b) Staining of Unfixed Bacteria

Exposure to solutions of basic dyes (neutral red, methylene blue, methyl green, azure, acetocarmine) of living bacteria usually does not result in the staining of nuclear material. The cytoplasm accepts the dye while the nucleoids stand out as clear areas. Later on, the bacteria become uniformly stained with the polyphosphate granules, when present, standing out as deeply stained bodies.

Fluorescent dyes such as acridine orange, acridine yellow, and auramine were used by several authors (SCHULER 1952, FLEGEL 1953, KRIEG 1954 a–d, FLOETHMANN 1954, KUSHNAREV 1959, MALATJAN 1963), but were found to damage the cells rapidly, especially in the presence of light, so that pathological effects affecting the shape and arrangement of nuclear bodies occurred (GIESBRECHT 1957). In several instances, acridine orange was incorporated in the growth medium in very low concentrations (10^{-5} or 10^{-6}). The cells were investigated in impression smears in water or a similar solution of the dye while some pressure was exerted on the coverslip of the mounted preparation in order to reduce the thickness of the liquid layer and background fluorescence, as well as to flatten the cells for easier observation (SCHULER, KRIEG). Artifacts caused by the dye appear in the form of an axial or spiral-shaped nuclear structure which may arise from several individual nuclear bodies by fusion (SCHULER 1952, GIESBRECHT 1957). Polyphosphate granules tend to accept the fluorescent dyes and therefore may cause interference.

c) Staining of Fixed Preparations

Fixation. As indicated above, preparation of stained bacterial specimens for cytological investigations requires that drying of the cells be avoided throughout the entire procedure. Many structural details seen in so-called "diagnostic" specimens prepared by flame fixation and a staining procedure involving repeated drying are artifacts, the usefulness of which for diagnostic purposes is unquestioned inasmuch as they can be obtained in a reproducible manner.

For cytological examination, usually a suspension of bacteria is spread onto an agar surface. The suspension should be dilute as to leave ample space between cells. From this agar surface impression smears are made with clean coverslips either immediately or after some period of growth. The coverslip is then transferred without delay to the container with fixative. Some authors grow or spread their bacteria onto thin agar layers on slides or in Petri dishes. For preparation, squares are cut out of the agar and placed on coverslips upside down. The coverslips are placed into the fixative with the agar in place. The advantages of this modification are probably minor.

The most commonly used fixative for nuclear studies is osmium tetroxide as a vapor. The smears are exposed to the vapors in a closed weighing glass with ground-on lid. The osmium tetroxide solution is one or two per cent and may be stabilized by addition of two to five per cent (w/v) of potassium dichromate. Exposure time should be limited to one or two minutes with bacteria of ordinary size. Extended exposure tends to produce artifacts in the form of rounded nuclear bodies (SMITH 1950, WINKLER and KNOCH 1951, GUHA et al. 1954, GIESBRECHT 1957). ELLER and BECKERT (1965) recommend a short exposure to nitrogen dioxide as a substitute for osmium tetroxide vapor.

Fixatives on the basis of ethanol and formalin are highly recommended by several authors. The Chabaud fixative which consists of 80 per cent ethanol (60 ml), crystallized phenol (15 g), 40 per cent formaldehyde (5 ml) and glacial acetic acid (2 ml) is recommended for studies with nuclease methods (BOIVIN et al. 1947 b), but the mixtures by Carnoy or Murray (the latter consisting of methanol with 1/10 volume of 40 per cent formaldehyde, MURRAY 1953) may be used with equal success. Mixtures containing mercuric chloride are suitable for use with the Feulgen nuclear reaction. The use of dichromate and picric acid fixatives has met with little success, and some earlier favorable results could not be verified by later investigators (e.g. MINCK et al. 1950, GIESBRECHT 1957).

Staining. Most bacteria, if treated with solutions of basic dyes, stain uniformly due to the presence of cytoplasmic RNA. Spores and lipid granules remain unstained whereas polyphosphate granules, if present, stand out as deeply stained structures. In cells with very little cytoplasmic RNA partial success may be obtained by the use of dilute solutions of the dye or by using the Giemsa stain (1 : 60 in pH 7.2 phosphate buffer) which will differentiate between DNA (reddish violet) and cytoplasmic RNA (blue) (JACOBSON and WEBB 1952).

The latter technique was modified to permit direct differentiation of nucleoids from other cell inclusions (except polyphosphates) in various types of bacteria. The cells are extensively stained in methylene blue, azure, or Giemsa, and subsequently treated with a 0.5 to 1 per cent solution of an acid dye (orange Y, eosine) for a very short period (one or two seconds). Several authors describe individually compounded mixtures of methylene blue eosinates for direct, progressive staining. These mixtures usually require preliminary experiments with cells of the particular strain to obtain a mixture which is best suited for the material under investigation. If these difficulties can be overcome, the results are probably the best that can be obtained with any light microscopic staining technique (Fig. 16).

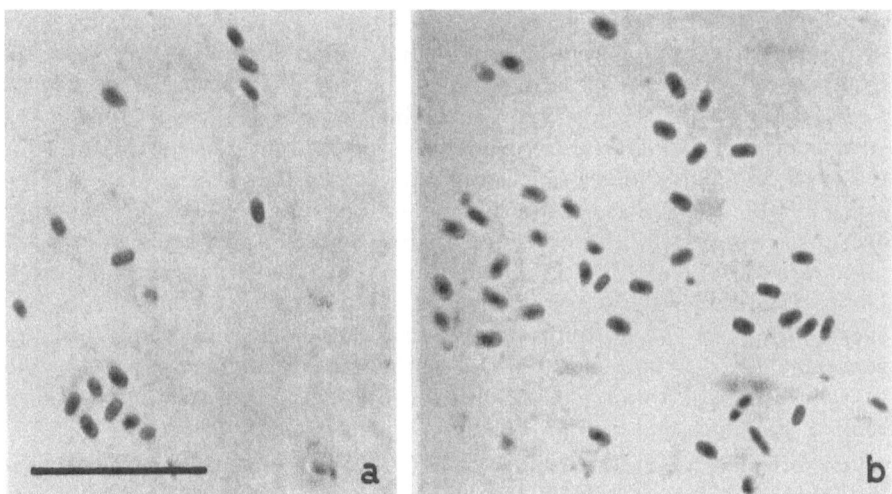

Fig. 16. Application of Piéchaud's stain to *Bacillus subtilis* germinating from spores. — *a* Grown for 30 minutes on nutrient agar at 37° C. — *b* Grown for 45 minutes. — Scale marker represents 10 μ. Photographs by Miss JENNIFER M. ALLEN, reprinted from BRIEGER (1963) with permission, Academic Press, Inc.

For details the reader is referred to ZETTNOW (1899), BADIAN (1930, 1933), PIEKARSKI (1937), TRONNIER (1953), HARTMAN and PAYNE (1954), and particularly PIÉCHAUD (1951, 1954, see also modification by BRIEGER 1963).

In other methods, special measures are taken to eliminate, prior to staining, interfering materials such as cytoplasmic RNA and granular polyphosphates. The best known of such preliminary measures is a hydrolysis in 1-N hydrochloric acid at 60° C for a period of four to ten minutes. This treatment results in the removal of RNA (VENDRELY and LIPARDY 1946), and after seven minutes of hydrolysis also in the removal of granular polyphosphates. DNA is the only basophilic cell constituent to resist such treatment, although it is physically and chemically altered. The removal of cytoplasmic material by acid hydrolysis is visible in the phase contrast microscope. The nuclear bodies appear as denser areas embedded in a less dense cytoplasm (PREUNER 1951, GIESBRECHT 1957).

After hydrolysis, nuclear staining can be accomplished with any basic dye. Examples are the Giemsa stain (PIEKARSKI 1937, 1939, ROBINOW 1942), basic fuchsin (von Plotho 1940, Smith 1950, Cassel 1950), crystal violet (Chance 1952), methylene blue, safranin and thionin. Azure or thionin with addition of thionyl chloride was recommended by DeLamater (1951), Huebschmann (1952) as well as by Robinow and Hannay (1953). Contrast obtained with this dye is excellent. Speculations concerning the mechanism and specificity of this method (DeLamater 1953) were never proved nor disproved. Quersin (1950) recommends the Gram stain after hydrolysis in acid. While direct staining of nucleic acids with methyl green and pyronin does not work with bacteria as it does with eucaryotic cells, pyronin alone may be used after hydrolysis and will stain the nuclear bodies (Tronnier 1952).

The only but extremely serious disadvantage of acid hydrolysis is the unsatisfactory preservation of fine structure. During hydrolysis, DNA undergoes partial depolymerization and complete denaturation which results in a drastic decrease in viscosity of the DNA-plasm. The result is a general tendency of the nuclear bodies to round off, and for narrow strands of nuclear material to disintegrate into fragments which will assume the shape of small round granules or droplets. If hydrolysis is restricted to a few minutes only, intermediate forms such as dumbbell-shaped structures are quite frequent, while extended hydrolysis results in a loss of nuclear material as apparent from a reduction in size of the (then spherical) nuclear bodies (Hoffman 1951, Giesbrecht 1957). Giesbrecht recommends a very short hydrolysis for optimum preservation of detail and points out that sufficient contrast may be obtained without completely eliminating cyto-plasmic RNA. It is understood, however, that such short treatment will not remove polyphosphate granules that may be present.

Historically, hydrolysis-basic dye techniques emerged as modifications of the Feulgen nuclear reaction. The basic dyes were used to improve the contrast between the nuclear bodies and the unstained background, and the similar-ity of the staining effects obtained with both techniques proved the specif-icity for DNA of the hydrolysis-basic dye methods. Actually the function of acid hydrolysis as a part of the Feulgen reaction is not to remove inter-fering basophilic material but primarily to prepare the DNA for the subsequent reaction with leucofuchsin by partially degrading it at the nucleotide level (Feulgen 1918, Feulgen and Rossenbeck 1924, Feulgen 1939, Stacey 1953, Lessler 1953). This is of importance with respect to the preservation of structural detail: hydrolysis time resulting in optimum color development in the Feulgen reaction is such that the formation of artifacts as described above cannot be avoided. Neither does the Feulgen reaction offer any advantage in the presence of polyphosphate granules, since the granules stain in a brilliant red with leucofuchsin alone and con-sequently must be removed completely by hydrolysis prior to the applica-tion of the leucofuchsin reagent. Since time of hydrolysis required to eliminate the granules (7 minutes) occasionally exceeds that required for optimum color development in the Feulgen reaction (5 to 8 minutes depend-ing on fixative and species), the granules, although Feulgen-negative, may interfere with the reaction (Reichenow 1928, Sassuchin 1935, Petter 1933 b).

Some substitutes for hydrochloric acid have been recommended in the literature but will not be dealt with in any detail (perchloric and trichloroacetic acids, CASSEL 1950, VENDRELY 1953; acid hydrolysis at reduced temperatures, STILLE 1937; alkaline hydrolysis, NERMUT 1959; hydrolysis with disodium phosphate, WELSCH and NIHOUL 1948, with snake venom diesterase, BOQUEL and LEHOULT 1948). It appears in general that the hydrolytic techniques mentioned above have outlived their usefulness as tools of cytological research and may be considered for exploratory or reference purposes only. The HCl-Giemsa technique with minimum exposure to acid may remain a valuable tool in elementary courses in bacterial morphology and cytology.

The use of enzymes for the removal of cytoplasmic RNA as indicated in the preceding paragraph has resulted in a considerable improvement. Ribonuclease isolated from pancreas was introduced into bacterial cytology by BOIVIN et al. (1947 a, b, 1948, 1949). The enzyme is used in a 0.02 per cent solution in a pH 6 phosphate buffer at 37° C; a 30 minute treatment is usually sufficient. Suitable fixatives are the Chabaud and methanol-formalin mixtures, a very short osmium fixation may be used if followed by intense rinsings with water. Giemsa stain in a 1 : 60 dilution in pH 7.2 phosphate buffer (staining time: two hours) may be used for the demonstration of the nuclear bodies. Contrast obtained with ribonuclease in this manner is inferior to that obtainable with acid hydrolysis (GIESBRECHT 1957).

A definite improvement of the ribonuclease technique consists in treatment of the preparations, after ribonuclease digestion, with a pepsin solution (0.02 per cent in 0.01 N HCl, 30 minutes at 37° C, PETERS and WIGAND 1953, WIGAND and PETERS 1954 a). Pepsin considerably clears the picture and increases the brilliance of the stain (GIESBRECHT 1957) presumably by removing protein components. The low hydrogen ion concentration of the pepsin solution is quite helpful in removing smaller polyphosphate granules prior to staining.

Deoxyribonuclease (0.02 per cent in phosphate buffer, pH 7.3, containing 0.03 M Mg^{++}, BOIVIN et al. 1947 a, b, 1948, 1949, TULASNE et al. 1948 a, b, YUASA et al. 1955) is useful as a control system for the specificity of all types of nuclear stains. Deoxyribonuclease specifically eliminates DNA without affecting the basophilia caused by ribonucleic acid or polyphosphates (PETERS and WIGAND 1953, JACKSON and DESSAU 1955).

Trypsin was tested in several instances but was found to exhibit nuclease activities affecting RNA as well as DNA, leaving the polyphosphate granules as the only basophilic cell inclusions (PETERS and WIGAND 1953, see also BAUMANN-GRACE and TOMCSIK 1959). These findings may not apply to purified products that have become available in recent years.

Techniques involving the use of a mordant prior to staining (e.g. CHANCE 1951, 1952, 1954, 1957, 1958) are generally inferior in specificity and reliability, and their usefulness as judged from published results appears doubtful. The iron hematoxylin technique (Heidenhain) has given excellent results in the hands of many earlier workers but inconclusive results in the hands of others. This may largely be due to the fact that the result depends on the differentiation of an overstained specimen until the right degree of staining is obtained. In this way, the investigator will either be guided by prior knowledge of the correct result or by a preconceived idea of same.

Stained bacteria are best observed in water mounts. With labile stains such as Giemsa, a neutral buffer may be used, or the specimen may be stabilized by a short treatment with a 3 per cent solution of phosphomolybdic acid. Dehydration of specimens and inclusion in resins is discouraged for several reasons: Dehydration of bacteria, no matter how slow, results in a shrinking to approximately one half of the original diameter (FUHS 1958, 1964). Such shrinking is intolerable in objects of the size of bacteria. Second, organic solvents may produce severe artifacts as shown by HALE (1954). Third, media of high refractive index tend to render certain structural details invisible that are of great importance in the critical study of bacterial nuclear structures, in particular the cell outlines including indications of multicellularity, and refractile cell inclusions that may affect the intracellular arrangement of nuclear material.

Excessive shrinking during dehydration may be avoided by using the freeze-dehydrating technique by DELAMATER (1951), but it should be noted that lipid inclusions are removed during extraction with the organic solvent, and that subsequent inclusion in resins will render the cell septa and boundaries almost invisible.

The freeze-dehydration technique has been held responsible for the production of an artifact referred to as "mitotic figures" (see CLARK et al. 1953, BISSET 1952–1954), but GIESBRECHT (1957) in a comprehensive evaluation found that many critics of the concept of mitosis in bacteria had no better comprehension of the phenomenon than had its originators. The artifact consists of a breakdown of the nuclear bodies into granules which show irregular arrangements some of which have been arbitrarily identified with chromosomal arrangements in a mitotic cycle. According to GIESBRECHT, the artifact is produced as follows: Growth of *B. megaterium* and other bacteria on an enriched agar medium produced cells loaded with lipid (poly-β-hydroxybutyrate) inclusions which occupied much of the intracellular space. The nucleoids in these cells were present in their "open" or branched form (see next section for details), with their branches or extensions arranged between and around the lipid inclusions. This system was broken up into islands or droplets of DNA-plasm by acid hydrolysis. While these droplets were stained clearly with thionin-SO_2 as described above, the lipid inclusions were extracted with solvent in the freeze-dehydration step, and subsequent inclusion in resin made the recognition of the original spatial relationship of the various cell constituents even more difficult.

D. General Morphology of the Bacterial Nucleoid

In the introduction (p. 5), a nucleoid has been defined as the cytological structure corresponding to one genetic complement or genome in the state of rest or replication. According to evidence presented above according to which the genome at least in some bacteria is a single DNA structure, the nucleoid also should be represented by a single coherent cytological structure, e.g. a nuclear body.

It has also been pointed out that several nucleoids may coalesce to form a single nuclear body. The nucleoids which participate in the formation of such compound nuclear body cannot be identified at the light-microscopic level and even hardly so at the electron-microscopic level of investigation.

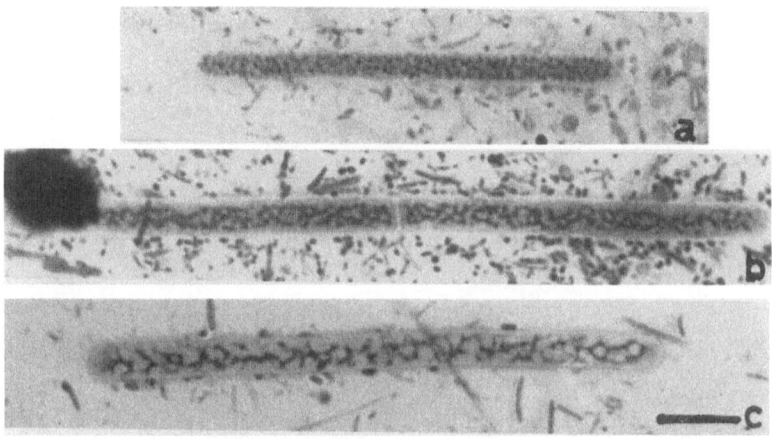

Fig. 17. Giant bacilli from intestine of tadpoles, osmium vapor fixation, Piéchaud nuclear stain. — *a* Common aspect of nuclear material: sponge-like reticulum, extending almost through all parts of the cell. — *b* Less frequent aspect of nuclear material consisting of a simpler network resembling Chinese characters. This cell is about to divide. — *c* Rare aspect of nuclear material consisting of a much simpler network. — Scale marker represents 10 μ. All photographs: Courtesy Dr. B. DELAPORTE.

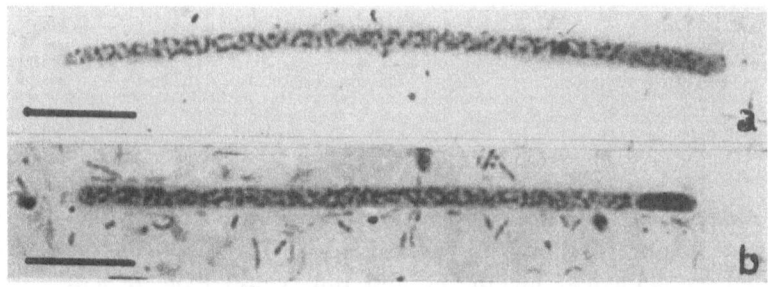

Fig. 18. Endospore formation in giant bacilli from the intestine of tadpoles. Osmium tetroxide vapor, Piéchaud nuclear stain. — *a Bacillus mirus* Delaporte. Part of the nuclear reticulum (shown at the right pole) has detached from the remainder and has somewhat changed its appearance. This part will enter the spore. — *b* More advanced stage of endospore formation in *B. mirus*. Details of nuclear structure in spore are recognized with difficulty. — Scale markers represent 10 μ. — Courtesy Dr. B. DELAPORTE.

One may consider also the opposite possibility, i.e. the distribution of nuclear material into nuclear bodies which are interconnected only by single or paired DNA duplexes. Such connections could not be detected in the light microscope and even would be hard to spot in ultra-thin sections. One may even consider the possibility that the entire genophor is dispersed evenly throughout the bacterial cell. Past experience shows, however, that neither of these two possibilities is materialized to any significant degree (see p. 84 for a possible exception). The DNA instead forms a "DNA

plasm" which is quite variable in shape, but even in its narrowest extensions
is made up of a sufficiently large number of DNA fibres to render such
extensions easily visible in the electron microscope and to make them—
at least in the larger bacteria—accessible to light-microscopic analysis.

A nucleoid which consists of a network or similar system of interconnected branches or strands of DNA-plasm will be referred to as a
nucleoid "in its open form". This term is used as a reminiscence

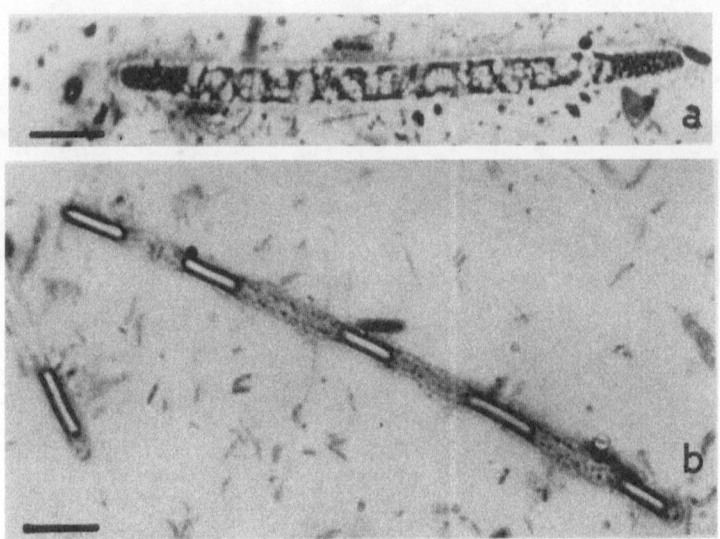

Fig. 19. Late stages of endospore formation in bisporal giant bacilli from the intestine of tadpoles. Osmium
tetroxide vapor, Piéchaud's nuclear stain. — a *Bacillus camptospora* Collin, bipolar endospore formation. Nuclear
reticulum in the forespores is quite dense. Intensity of nuclear stain in other parts of the cell varies. In the photograph, the most intensely stained areas appear in black, the moderately stained strands are gray while the most
delicate or lightly stained portions are not seen. — b *Bacillus speciosus* Delaporte, chains of cells containing two
mature spores each. The chain shown in the photograph consists of two entire cells as indicated by the transverse
walls, plus one cell containing one spore only. This spore is about to be liberated. The single spore shown in the
lower part of the photograph may be the other spore derived from that cell. Remnants of the chromatin material
are faintly indicated in the spaces between the spores. — Scale markers represent 10 μ. Photographs: Dr. B. DELA-
PORTE.

to Hieronymus' (1892) term "open nucleus" which was applied in exactly
the same manner and was' intended to describe a nucleus which was not
bounded by a nuclear membrane and thus could exist in a less compact
or reticular form. This configuration is also known as "chromatic reticulum"
or, in the French literature, as "réseau nucléaire", although in most cases,
the superposition of the strands of a moderately branched structure may
create the impression of a true network.

Lucid descriptions of the "open" nucleoid and the open form of the
compound nuclear body were given by many authors. The earlier findings
and a comprehensive review of the larger bacteria are found in publications
by Delaporte (1939, 1940). Excellent examples are also found in the giant
Bacillus- or *Spirillum*-like species (Delaporte 1964 a, b) (Figs. 17 and 18),
using newer methods. Open nuclear configurations in smaller bacteria were
found by phase-contrast observation of living cells (Mason and Powelson
1956, Robinow 1960) (Fig. 15). The structure of the open nucleoid is clearly

fibrillar, although granular components were described frequently. Their appearance can often be correlated with the use of acid hydrolysis which may result in the disruption of the finer fibres and the formation of tiny droplets (see above). This idea is supported by the relative absence of such granules in iron hematoxylin preparations. Polyphosphate granules may account for another part of the apparent granular components. In the larger bacteria, orientation perpendicular to the specimen plane of segments of the reticulum may create the impression of a granular component.

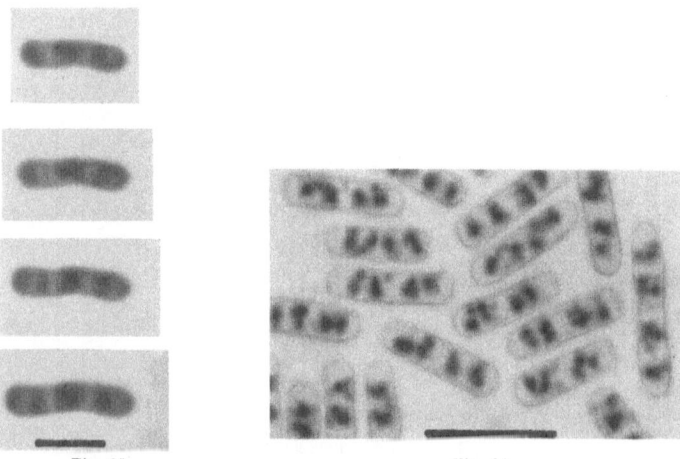

Fig. 20.   Fig. 21.

Fig. 20. *Achromobacter sp.*, sequence of phase-contrast micrographs of cell undergoing nuclear division in left half of cell. Nucleoids appear as transversely oriented bars. Scale marker represents 2 μ. Reprinted from ROBINOW (1956).

Fig. 21. *Bacillus cereus*, fixed in osmium tetroxide vapors, HCl-Giemsa nuclear stain. The cells show transversely oriented nucleoids and V-shaped division stages. Scale marker: 5 μ. Courtesy Dr. C. F. ROBINOW.

In many reports the appearance of the nucleoid is described as that of a t r a n s v e r s e l y o r i e n t e d b a r o r r o d (BADIAN 1933, ROBINOW 1944, KLIENEBERGER-NOBEL 1945, 1947 b, TULASNE and VENDRELY 1947 a, b, BISSET 1948 a, b, 1949, 1952 e, BERGENSEN 1953 a, WILLIAMS 1959) (Figs. 20 and 21). GIESBRECHT (1957) considers the nucleoid in *B. megaterium* to be a transversely oriented bar and presents evidence that prolonged exposure to osmium vapors or acid hydrolysis can convert such nucleoids into more rounded, dumbbell-shaped and even spherical structures. Transversely or obliquely oriented nucleoids have also been observed in the phase-contrast microscope (TULASNE 1949 a, b).

The s p h e r i c a l n u c l e o i d s described in many of the earlier publications in some instances must be considered an artifact or a product of poor resolution, the latter point being still valid for many small-celled bacteria. Electron microscopy (see below) is generally in favor of some or other prolate shape. Spherical nucleoids were described for resting cells of many bacteria (BISSET, see below), but specimens were usually prepared with the osmium fixation and acid hydrolysis techniques. For spherical nucleoids after chloromycetin treatment see p. 47, 48.

More extended nuclear structures located close to the cell membrane as a p e r i p h e r a l   s p i r a l   were observed in *Spirillum spp.* (WILLIAMS and RITTENBERG 1957, YUASA et al. 1958, TANAKA 1961) (Fig. 22). A similar structure is the horseshoe-shaped nucleoid in the narrow cellular compartments of *Caryophanon latum* (PESHKOFF 1946, PRINGSHEIM and ROBINOW 1946).

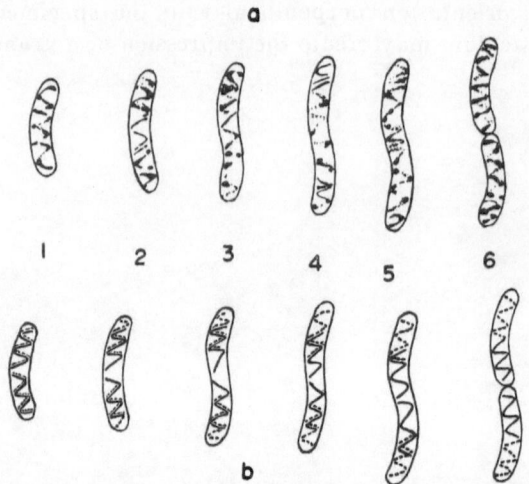

Fig. 22. Division of chromatin bodies in *Spirillum undula*, Strain S 30. — *a* Drawings from Feulgen-stained preparations arranged according to the length of a cell. — *b* Schematic drawings of the division of chromatin bodies. From TANAKA (1961).

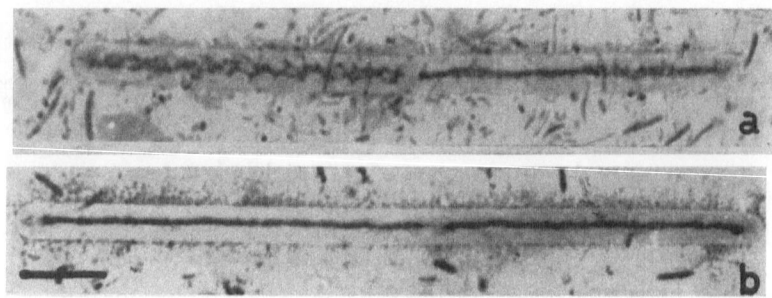

Fig. 23. Giant *Bacillus sp.* from intestine of tadpoles. Osmium vapor fixation, Piéchaud nuclear stain. Nuclear material more or less condensed into an axial filament. This aspect is quite rarely observed in these forms. Scale marker represents 10 μ. — Courtesy Dr. B. DELAPORTE.

A frequently observed structure is the n u c l e a r   a x i a l   f i l a m e n t (Fig. 23). DELAPORTE (1939/40) described this type of nuclear structure in bacteria which according to other authors have nucleoids in the shape of transverse rods. A possible explanation for this discrepancy is given below (p. 71). The axial filament type usually is observed in advanced phases of the cell cycle, i.e. in the beginning stationary phase or at the inception of spore or microcyst formation (for references see below, p. 82). Extended axial rods usually do not represent single nucleoids but are compound organelles containing several nucleoids each. For a correlation of light and electron microscopic findings on nucleoid morphology see p. 104.

E. Morphology of the Dividing Nucleoid

Despite the currently popular belief that the search for a valid pattern of nuclear division with the light microscope was a complete failure, there is sufficient indication that a few patterns have emerged from the wealth of published information providing insight into the division process. Reports describing preparations that are essentially free from technical imperfections and in which nucleoids are shown to be of the spiral or transverse-rod type emphasize that these structures divide longitudinally. In the case of transversely oriented structures, longitudinal splitting may occur producing a V-shaped nucleoid, whereupon the daughter structures separate completely and move apart (BADIAN 1930, 1933, TULASNE and VENDRELY 1947 a, b, phase-contrast observations by PESHKOFF 1946, TULASNE 1949 a, b, CLIFTON and EHRHARDT 1952). The latter authors describe the longitudinal splitting of transversely oriented rods and point out that if several nuclear divisions occur within one cell, the splitting process proceeds in the same direction in all nucleoids. Some investigations with the HCl-Giemsa technique have provided comparable results (ROBINOW 1944, 1949 a, b, and subsequent papers; BISSET, numerous publications, GIESBRECHT 1957).

The peripheral spirals in *Spirillum* divide by longitudinal fission. Daughter nucleoids separate by sliding alongside each other. These conclusions are based on estimates of length and thickness of the nuclear spiral as determined in stained preparations (TANAKA 1961, Fig. 22). WILLIAMS (1959) found transversely oriented bars, dumbbell- and V-shaped bodies to be the predominant nuclear configurations in the related species, *Spirillum annulus*.

Investigations with the improved phase-contrast technique revealed that shape and intracellular arrangement of nucleoids changes slowly but constantly throughout the entire division cycle (ROBINOW 1960, see Fig. 15). MASON and POWELSON (1956) also found that at a generation time of 30 minutes, 24 minutes were spent on nuclear replication while separation of the daughter structures took 6 to 8 minutes. Cell division was completed 12 minutes after nuclear division. These findings explain the apparent difficulties in establishing a simple and universal geometrical pattern of nucleoid division.

The most puzzling of the nuclear configurations, as far as nuclear division is concerned, is the axial filament which, according to light-microscopic observations, undergoes longitudinal growth followed by transverse fission. Electron microscopic analysis, however, suggests a divisional process not basically different from those described above (see p. 121).

F. Environmental Factors and the Morphology of Nucleoids

In the past, reference to the physiological state of a bacterium too often was a convenient way of dealing with unexplained differences between results of individual investigators. Only few authors studied the nuclear configurations in bacteria as related to well-defined external conditions. Moreover the frequent use of agar plate and poorly controlled liquid batch cultures has resulted in much confusion as to the environmental and developmental factors active during the bacterial growth cycle. In the following sections an attempt will be made to differentiate between these two aspects. The present section will deal with factors that most obviously are of the environmental type.

Most important in several respects is the finding that changes in the solute content of the medium can cause reversible changes in the appearance of the nucleoids. Johnson and Gray (1949) found that discrete, compact, centrally located nucleoids became more diffuse when the salt concentration of the medium was reduced. More thorough studies were undertaken by Whitfield and Murray (1956). Transfer of *Salmonella typhimurium* cells from salt-free medium into a medium containing one per cent sodium chloride caused a slow condensation of "open" nucleoids into axial bodies or filaments; transfer into 2 per cent NaCl medium resulted in a change of configuration within a few minutes. Recovery occurred within one hour after subsequent transfer into salt-free medium, provided that cell metabolic functions were still intact. The effect was observed with the chlorides and acetates of sodium, potassium, lithium, and calcium (Fig. 47, p. 105).

A shift in temperature from 37° to 4° C had a similar condensing effect on nuclear structure (Whitfield and Murray 1956). The effects of temperature and changes in salt content of the medium were synergistic. A shift to 4° C in the absence of salt at first caused aggregation of the nuclear material, then nuclear fragmentation and apparent loss of viability of the cell eventually followed by lysis (see also Klieneberger-Nobel 1947 c).

Other factors causing aggregation of nuclear material are starvation, increase of pH to a value of 10, exposure to 2,4-dinitrophenol or dihydro-streptomycin. (For streptomycin action see also Tulasne et al. 1948 b and Preuner and von Prittwitz 1951.)

Nermut (1962) found that the effect of solutes was not restricted to electrolytes. In his experiments with *B. megaterium* and *Proteus* strains, media containing $1 M$ (1.8 per cent) glucose caused nuclear aggregation. The effect of glucose resembled that of sodium, calcium or magnesium salts in a comparable range of concentrations. Cooling had some effect except in freshly prepared salt-free media. Nuclear aggregation was also observed in salt-free media that were in the process of drying out. Also Preusser (1958, 1959) observed a condensation of nucleoids on solid media.

Nuclear condensation occurs also upon exposure to air of anaerobic bacteria and upon exposure to hydrogen peroxide of anaerobic or aerobic bacteria (Klieneberger-Nobel 1945, Cassel 1951 b, Delaporte 1958, see also Minck and Minck 1948).

Ultraviolet irradiation can cause condensation of bacterial nuclear structures (Payne et al. 1956), but this effect is transitory and salt-dependent (Whitfield and Murray 1956). The ultimate effect of ultraviolet light consists in the fragmentation of the nucleoid and cell death (Kellenberger 1952 a).

Colchicin was found not to affect nuclear configuration or division in bacteria. The effects of various other agents were studied, but the significance and specificity of the observed changes is uncertain (PARVIS 1954, DELAMATER et al. 1955, SCHWEISFURTH 1959).

Findings reported in this section shed light on several facets of bacterial caryology and offer explanations for several discrepancies that are evident for anyone reviewing the literature in this field. DELAPORTE's frequent descriptions of axial nuclear filaments are correlated with her preference for media containing 2 per cent glucose. Nuclear fusion was consistently described in ageing cultures of various bacteria by BISSET and others but cannot be clearly separated from the environmental effects described here and is questionable as a developmental phenomenon.

G. Developmental Processes

1. Criteria

The criteria for many "developmental cycles" as described in the cytological literature are vague and superficial. Many of the changes observed parallel environmental changes that occur in agar and liquid batch culture media during the bacterial growth cycle. Changes involved are not only the supply and the exhaustion of the substrate which cause growth and cessation of growth in bacteria (metabolic "upshift" and "downshift"). Other and less well defined changes may occur simultaneously and may equally affect growth and cell structures, e.g., accumulation of metabolites, drying out of the agar surface.

Environmental effects on cell structure comprise such effects that are reversed upon alleviation of the external condition. In many cases, the initial intracellular event caused by the external condition in turn may initiate a sequence of secondary events (as is the case e.g. in the synthesis of macromolecular components during metabolic upshift). In that case our definition would require that after the reversal of the initial step the sequence is interrupted and the cell returns to its original condition.

A developmental cycle would also involve an initial change from condition "A" to condition "B" triggered externally, but upon alleviation of the external condition, the cell would not return to "A" but due to the operation of an autonomous mechanism of control would pass through at least one more phase ("C") before returning to condition "A".

Since in this section we are not concerned with autonomous mechanisms on the molecular level of organization, it appears that the nuclear cycle of many bacteria is entirely controlled by environmental factors.

In the absence of any conclusive proof of the contrary this seems also to apply to the phenomenon of nuclear aggregation or fusion as reported by many authors in a variety of bacteria (KLIENEBERGER-NOBEL 1945, 1947 b, c, BRIEGER and ROBINOW 1947, BISSET 1948 a, b, c, 1949, BISSET et al. 1951, GRACE 1951, WEBB et al. 1954, HOPWOOD and GLAUERT 1960 a, b, SHAMINA 1964). Upon exposure of bacteria to monofluoroacetate, the fusion of nuclear bodies could be observed in the phase-contrast microscope (HIRANO 1961). The effect may be related to changes in humidity on agar plates or other less specific conditions such as exhaustion of nutrient supply in ageing cultures. Its interpretation as a meiotic event (Bisset et al. 1951, Hirano 1959, Prévot and Mazurek 1953) is not supported by any specific evidence and shows little consideration of the haploid nature and the mode of nuclear segregation in bacteria. Moreover, the apparent ease of coalescence and separation of nuclear bodies representing individual nucleoids advises against overestimating the genetical or physiological significance of such phenomena.

2. Events Suggestive of Genetic Processes

According to Bisset (1948 d), the nuclear bodies in adjacent cells of multicellular bacteria may undergo apparent fusion. The significance of this phenomenon which was observed in R-forms of *Lactobacillus sp.* is not clear, and some of the arguments presented in the preceding sections may apply.

In *Spirillum spp.*, two organisms can become entangled and may undergo longitudinal fusion. Nucleoids of both individuals become oriented opposite each other and undergo fusion after cell boundaries have been dissolved (Williams and Rittenberg 1957). In the absence of other supporting evidence the authors are careful in not interpreting their observations in terms of a sexual process.

Angular association and subsequent fusion of cells and nuclear bodies in *Bacterium malveolarum* bear some resemblance with the events occurring in *Spirillum* (Stoughton 1932). Stapp (1942) who earlier discounted Stoughton's observations as artifacts, found aggregation of cells in star-like formations in various strains of *Pseudomonas (Agrobacterium) tumefaciens* and related bacteria (see also Stapp and Knösel 1954, 1956 a, b, c). Star formation was accompanied by the migration of the nuclear bodies towards the centrally located cell poles. The close association of cell poles and nuclear bodies in the light microscope is virtually indistinguishable from fusion of the nuclear material into a single DNA plasm. Critical cytological and genetical examinations, however, failed to reveal multiple cellular and nuclear fusion although contact between adjacent cells was intimate (Heumann 1963, Marx and Heumann 1963, Heumann and Marx 1964). Pili were instrumental in establishing cell-to-cell contact. The authors found an exchange of genetic material, but contrary to earlier speculations the exchange involved only single pairs of cells within the star-like formation (p. 43).

"Fusion tubes" in *Bacillus* species (DeLamater) were found to be rows of small cells adjacent to normal cells (Bisset 1952 b). Such small cells are occasionally formed on suboptimal media together with cells of normal size.

While the cytological literature contains numerous speculations on sexual phenomena without providing conclusive evidence, cytological observations of established processes of genetical exchange are extremely scarce. Electron-microscopic observations on the process of bacterial conjugation are cited earlier in this treatise (p. 39); also microscopic indication of DNA excretion has been mentioned (p. 44). Episomes thus far have not been identified with either the light or electron microscope (Kellenberger and Kellenberger 1956); the size of the known episomes defies light-microscopic identification.

3. Endospore Formation

Endospore formation in bacteria is a complicated sequence of events partly controlled by internal factors. A complete presentation including physiological aspects (SCHAEFFER et al. 1965, VINTER and SLEPECKY 1965, REMSEN et al. 1966, SRINIVASAN 1966 and others) as well as genetical aspects (e.g. RYTER et al. 1966 a, b) exceeds the scope of this review. A critical

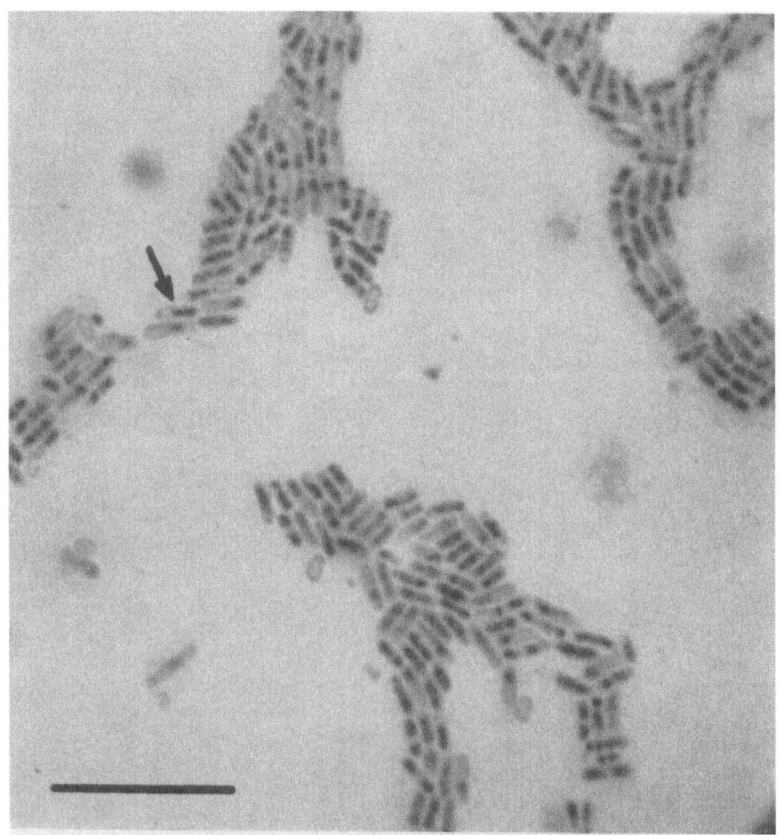

Fig. 24. The formation of spores in *Bacillus subtilis*. Piéchaud's nuclear stain. Note apparent fusion of chromatinic body at arrow. Scale marker represents 10 μ. Photograph by Miss J. M. ALLEN, reprinted from BRIEGER (1963) with permission, Academic Press Inc.

phase in the completion of the spore is reached when the forespore "buds off" into the cytoplasm, after which event the entire maturing spore is separated from the environment by a layer of cytoplasm and two complete membrane systems. Further maturation will not occur if the cells were converted to protoplasts at an earlier stage (FITZ-JAMES 1960, 1964 a).

The spore DNA is identical with the DNA of the vegetative cells in its physical conformation and in its base composition (MANDEL and ROWLEY 1963), but its quantity is reduced with respect to the amount of DNA in the vegetative cell, in that only one genetic complement (i.e., one nucleoid) is incorporated into the spore. This is evident from chemical analyses (FITZ-

James 1955 a, Bennett and Williams 1960), from determinations of sensitivity against radiation (Lea 1955, Donnellan and Horowitz 1957, Woese 1958) and has been inferred from cytological observations (Badian 1933,

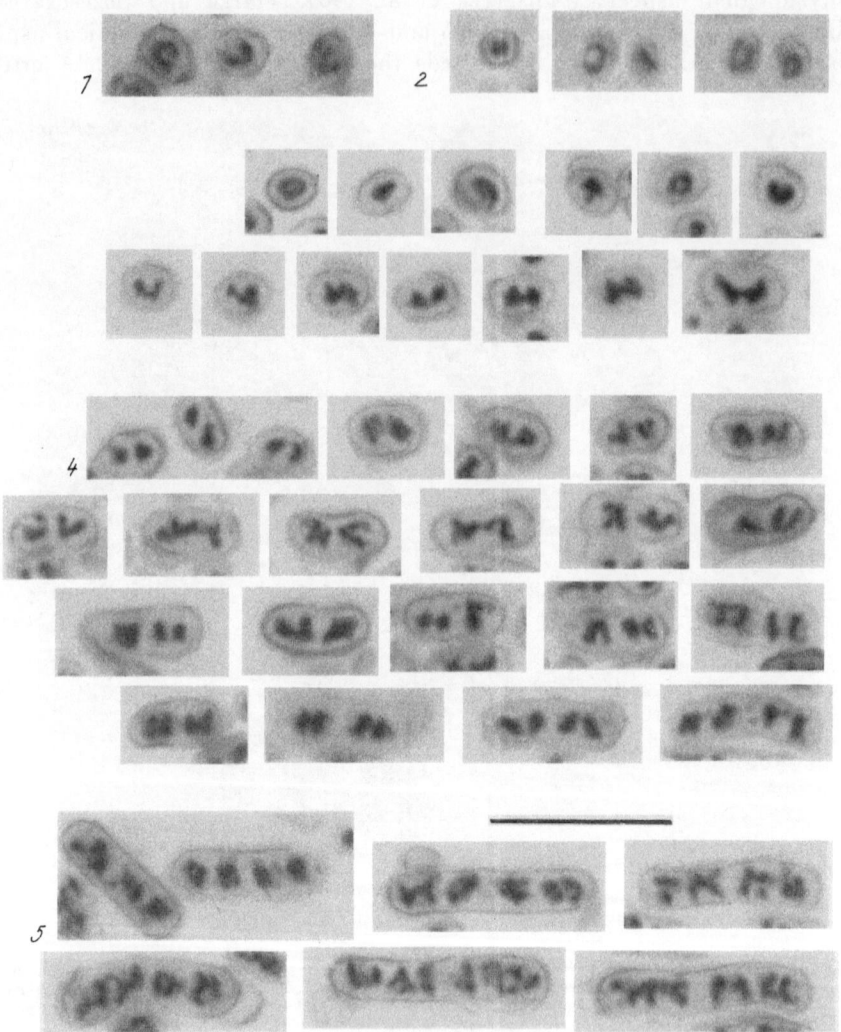

Fig. 25. Sequence of phases of spore germination and early growth of *Bacillus megaterium*, as demonstrated by the HCl-Giemsa staining technique (osmium vapor fixation). Sequence marked "3" ends with the formation of two nucleoids per cell, sequence marked "4" shows transition from two- to four-nucleated state. Sequence marked "5" shows loss of synchrony in subsequent growth. — Scale marker indicates 10 μ. Courtesy Dr. C. F. Robinow.

Piekarski 1940, Flewett 1948, Rolly 1951). As a species-related character, a polar or a more centrally located nucleoid may become the spore nucleoid.

In many instances, the appearance of an individual spore nucleoid is preceded by the apparent fusion of the DNA-plasm into a single axial filament as described previously (see also Ryter 1958), but some observers emphasize the non-essentiality of this phase (e.g. Flewett 1948). If a compound nuclear body is formed in this manner, the spore nucleoid emerges as a polar or central portion of that compound nuclear structure. In some

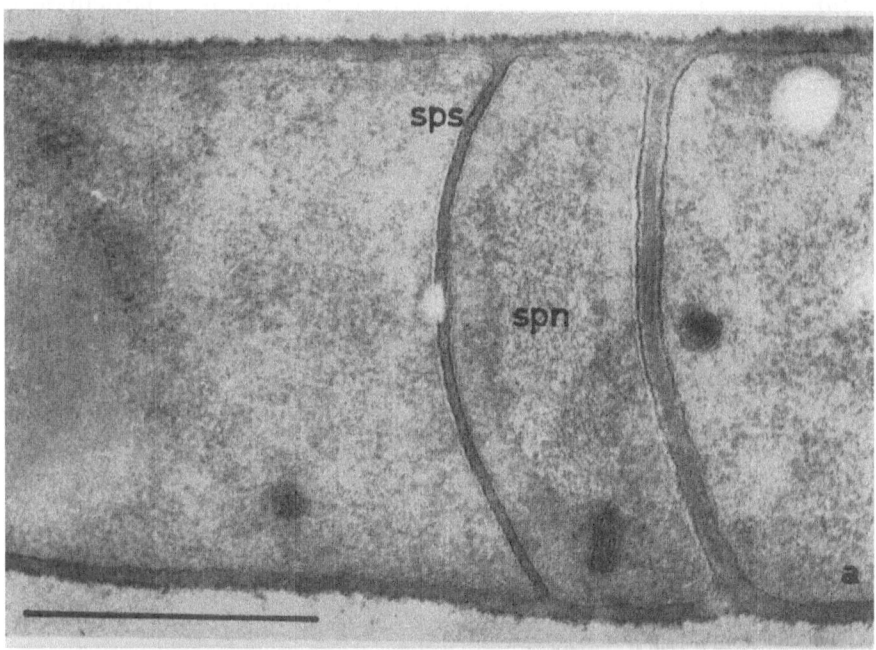

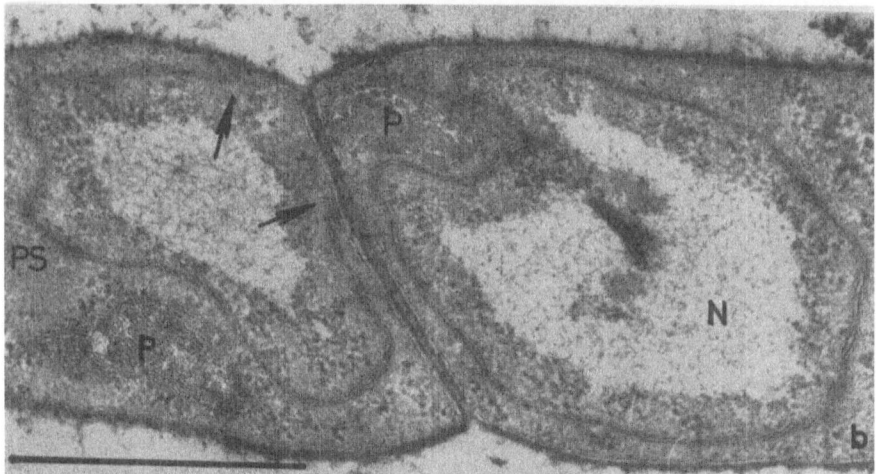

Fig. 26. The fate of the nucleoid during spore formation in *Bacillus cereus* var. *alesti*, ultrathin sections. — *a* Boundary between two cells of a chain indicated by transverse septum at the right. A spore septum (*sps*) consisting of cytoplasmic membrane material has grown in and has created a compartment containing a nucleoid (*spn*) which later will become the nucleoid of the spore. A plasmalemmosome is shown which is in contact with the nucleoid. The plasmalemmosome is connected to the membrane above or below the plane of sectioning. — *b* Two phases of forespore formation seen in two adjoining cells. In the left cell formation of forespore membrane is still incomplete (arrows), in the right cell the forespore is completely surrounded by its membrane. Nucleoid in right cell is marked "*N*", "*P*" are plasmalemmosomes, "*PS*" denotes a parasporal body. Scale marker is 0.5 μ. Courtesy Dr. P. C. Fitz-James.

instances, the compound nuclear body will again separate into individual nucleoids, one of which becomes the spore nucleoid (KLIENEBERGER-NOBEL 1945).

Nuclear structures of the sporeforming cell which do not become incorporated into the spore are degraded while the spore matures. In *Bacillus subtilis* it was found that all nucleoids of the sporangium participate in the synthesis of messenger RNA specific for the sporulation process, but activity of nucleoids other than the spore nucleoid ceases some time during

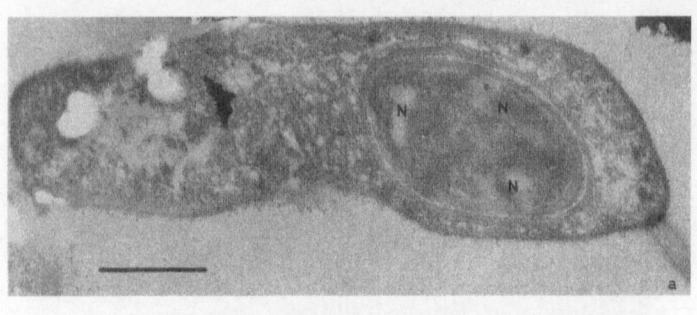

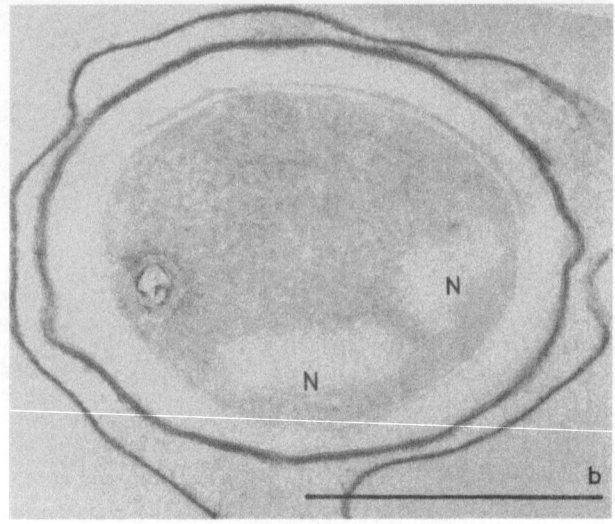

Fig. 27. Mature spores of *Bacillus sp.* — *a* Spore not yet liberated from cell shows vague outlines of its peripheral band-like nucleoid (*N*), nucleoid remaining in cell is in process of degradation. — *b* Liberated spore, nucleoid as in *a*. No structural detail of spore nucleoid can be seen which presumably is due to the difficulty of adequate fixation through the spore coat. — Scale markers represent 0.5 μ. — Courtesy Dr. P. C. Fitz-James.

spore maturation (Ryter et al. 1966a). In bisporal species (Delaporte 1964a) (Fig. 19) the situation is similar in that only part of the DNA is incorporated in the spores. *Metabacterium polyspora* (Robinow 1957) is the only bacterium known which uses its entire chromatin to form numerous endospores.

Sporulation in *Clostridium* species appears to resemble closely that in *Bacillus* (Hashimoto and Naylor 1958, Takagi et al. 1960, Robinow 1960, Hoeniger and Stuart 1967).

The appearance of the nucleoid in the mature spore was controversial for a long time. This was largely due to the difficulty of staining the spore nucleoid with techniques not involving acid hydrolysis. Acid hydrolysis,

however, may result in extrusion of the nucleoid from the spore or from
the interior of the spore into spaces between layers of the spore wall
(PREUNER 1951, 1953, ROBINOW 1953 a, b, FITZ-JAMES et al. 1954). The effect
is believed to be related to the history of the spore, in particular to the
medium on which the sporangium was grown (HUNTER and DELAMATER
1955).

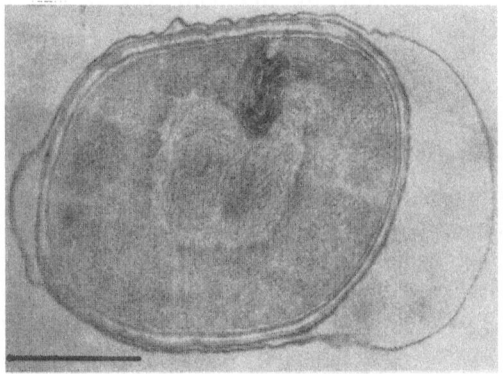

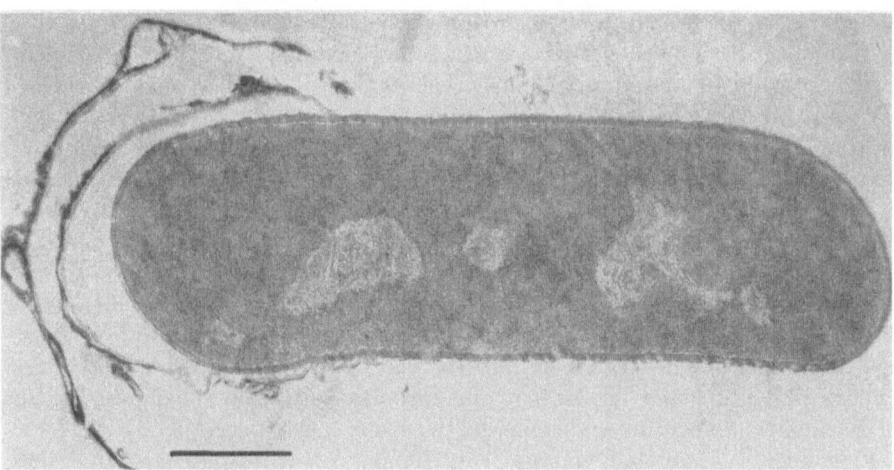

Fig. 28. Spore germination in *Bacillus sp.* — *a* Early phase of germination, nuclear detail becomes visible. Nucleoid has been converted into a centrally located, compact structure. — *b* Late phase of germination after nuclear division. Cell is shedding spore coat. — Scale markers: 0.5 µ. — Fig. 28 *a*: Courtesy Dr. P. C. FITZ-JAMES. Fig. 28*b*: Courtesy Dr. C. F. ROBINOW.

The normal spore nucleoid is oblong, crescent- or horseshoe-shaped or
branched and situated at the periphery of the spore (Fig. 27). The spore
cytoplasm is distinctly basophilic and appears opaque in the ultraviolet
microscope (ROBINOW 1953 a, b, FITZ-JAMES 1953, HASHIMOTO and GERHARDT 1960).

During germination of spores, DNA synthesis is initiated, but the first
cycles of replication do not result in cell divisions but serve to bring the
DNA content of the young cell to the level appropriate for the respective
physiological state (FITZ-JAMES 1954, 1955 a, b, STUY 1958, YOUNG and FITZ-
JAMES 1959).

Demonstrations of sporulation and of spore nucleoids in the light microscope are not particularly impressive due to the methodological difficulties
involved (Fig. 24). An excellent light micrograph of sporulation in
Clostridium sp. was presented by Robinow (1960). For sporulation in giant
bacilli see Figs. 18 and 19. Phases of spore germination are depicted in
Figs. 15, 16, and 25. The electron microscope is a more adequate means of
investigation of the morphology of nucleoids during spore formation and
germination as well as for the demonstration of the nucleoid in mature
spores (Figs. 26, 27, 28, and 58). An excellent sequence of illustrations was
also presented by Kawata et al. (1963).

4. Conidia Formation and the Life Cycle of the Actinomycetales

The tendency towards multinuclearity and the formation of multicellular filaments during phases of active growth and subsequent change to an
uninuclear short-celled condition upon cessation of active growth is quite
well expressed in various members of the *Actinomycetales.* In *Nocardia,*
multinuclear filaments persist during active growth and disintegrate into
uninuclear cells when entering the stationary phase (Salvatore and
Pontieri 1956, Hagedorn 1959 a, b, Adams 1963). Other forms develop
c o n i d i a by fragmentation of more or less specialized sections of the
mycelium. Such conidia are characterized by a special wall that is formed
inside the wall of the sporogenous mycelium with the latter temporarily
serving as an outer sheath. The fragmentation of the outer wall results in
the liberation of the conidia. Upon germination, the wall of the conidium
becomes the wall of the new filament (Glauert and Hopwood 1961) (Fig. 29).

In *Streptomyces* species, formation of conidia is restricted to hyphae
of the aerial mycelium which arises as a secondary mycelium from the
mycelium that penetrates the growth substrate. The primary or substrate
mycelium is reported to develop swellings containing several nucleoids each.
These formations are the origin of the secondary mycelium (McGregor
1954). In other forms, fusion of hyphae has been assumed to occur in the
primary mycelium and to result in the formation of basal cells from which
the secondary mycelium may emerge (Klieneberger-Nobel 1947 a).

Fusion of hyphae that emerge from separate colonies is considered the
most likely mechanism for the formation of heterocaryons, i.e. mycelia with
genetically diverse nucleoids operating in a common cytoplasm (Szybalski
and Braendle 1956, Braendle and Szybalski 1957 a, b, Bradley and Lederberg
1956 a, b, Bradley 1957, Horvath 1963). There are, however, more combinations of strains that undergo hyphal fusion than combinations that give
rise to heterocaryons ("compatibility", Bradley et al. 1959). This may be
due to the fact that cooperation as well as an approximately equal rate of
replication of the two types of nucleoids is required to maintain the heterocaryotic state; also a certain pattern of distribution of daughter nucleoids
within the mycelium is required to render the heterocaryotic character a
hereditary property of minute segments of the mycelium, in that
such segments must harbor at least one nucleoid of either type.
The pattern of distribution of genetically different nucleoids in

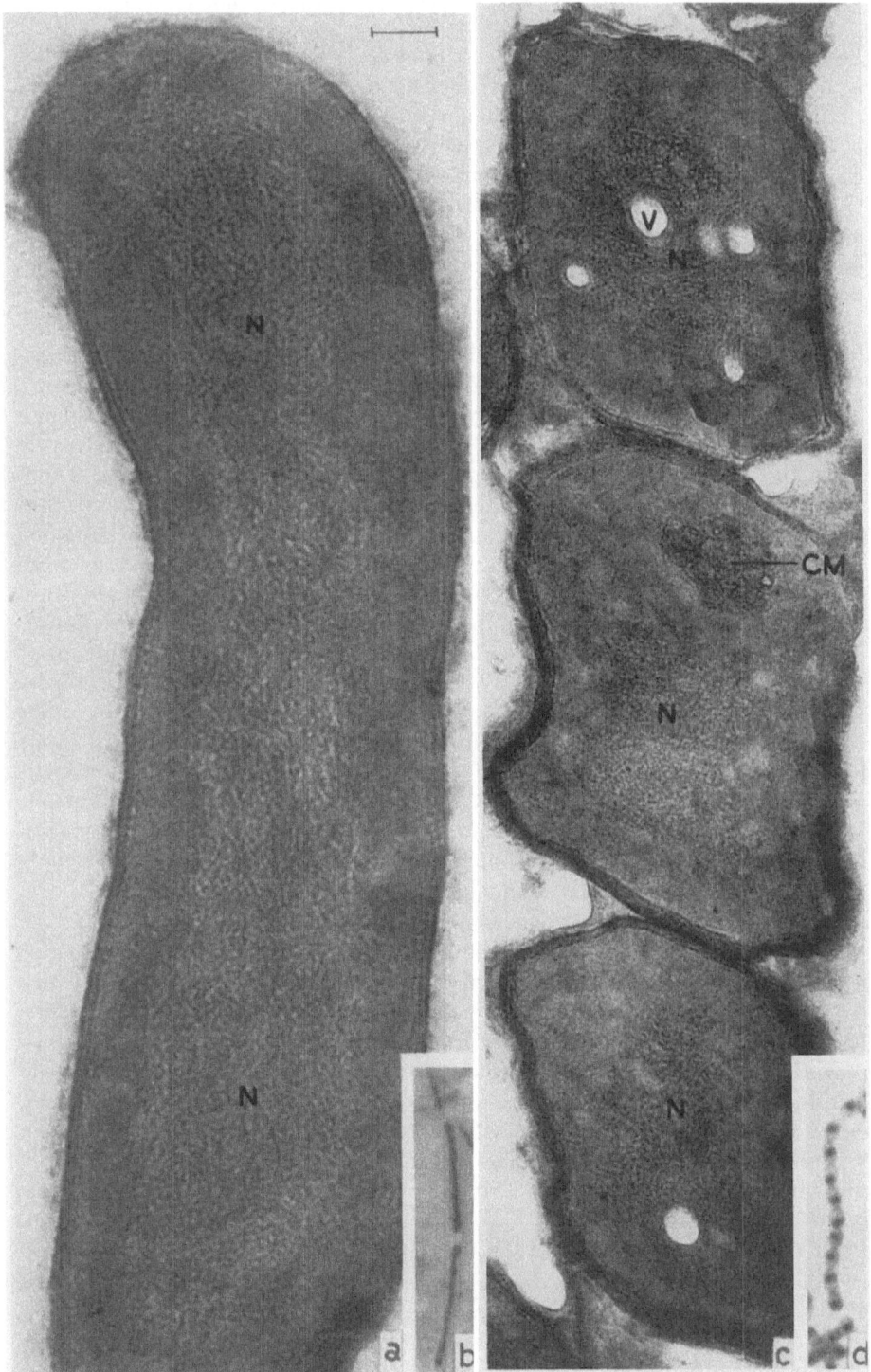

Fig. 29. Light- and electron-microscopic photographs of conidia formation in *Streptomyces coelicolor*. — *a* and *b*, Young aerial hypha from 48 hours old colony on minimal medium. Axial strand of nuclear material containing several genomes. The light micrograph also shows the formation of the conidia which contain one nuclear body each. — *b* and *c*, chains of mature conidia. — Scale marker 3.3 μ for light micrographs, 0.1 μ for electron micrographs. Reprinted from HOPWOOD and GLAUERT (1960 b).

heterocaryotic prototrophs was studied by micromanipulation of hyphal segments as well as by analyzing the progeny derived from single chains of conidia. The conidial chains of some strains were found to contain a specific parental genome only, while the conidia of others transferred characters inherent to either parental genome. It was concluded that the chains were formed by simple fragmentation of the aerial hyphae (BRAENDLE and SZYBALSKI 1959, BRADLEY 1959). Parental segregants from heterocaryotic strains showed a much greater tendency to combine again than the original parental strains had shown before. This tendency was transmitted through conidia often enough to be considered a new hereditary trait but later was lost in an unpredictable manner (BRAENDLE and SZYBALSKI 1959).

The nucleoids in heterocaryotic strains were found to participate in an exchange of genetical characters, presumably by recombination. Although the rareness of this event clearly renders BISSET's ideas of a sexual cycle untenable, the process as such is puzzling since it cannot be explained in terms of the known processes of genetic exchange in bacteria (SERMONTI and SPADA-SERMONTI 1955, 1959, BRAENDLE and SZYBALSKI 1957 a, b, 1959, HOPWOOD 1959). On the other hand, one may argue that bacterial or related genomes or fractions thereof embark in genetic recombination, as soon as they are incorporated in a common cytoplasm. The heterocaryon therefore represents a system in which such events are possible, and the low frequency of their occurrence rules out excessive diploidization as a possible mechanism. HOPWOOD's findings of reciprocal recombinants were not confirmed by other workers who prefer models involving merozygotes as found in other systems of genetic exchange in bacteria. The occurrence of oblong or dumbbell-shaped nucleoids in the secondary mycelia (VON PLOTHO 1940 and others) was generally assumed to be indicative of diploidization, but HOPWOOD (1960) in phase-contrast microscopic investigations found that such nucleoids did not arise from a fusion of nuclear bodies. (And, as pointed out above, even fusion would not represent conclusive proof of diploidization.)

The conidia of several members of the genus *Streptomyces* contain one nucleoid only. Light-microscopic preparations show a single nuclear body (Fig. 29). A one-hit survival curve is obtained in irradiation experiments. In these strains, conidia cannot transmit the character of heterocaryosis. A smaller number of strains was found to produce binucleated conidia. Single conidia from such strains may produce heterocaryotic mycelia. The survival curve follows a two-hit pattern. Cytological investigation does not always reveal the binucleated state. The conidial nuclear body may appear as dumbbell-shaped, oblong, or simply round (VON PLOTHO 1940, KLIENEBERGER-NOBEL 1947 a, HAGEDORN 1955, KINOSHITA and ITAGAKI 1958, SAITO and IKEDA 1957, 1958, 1959, HOPWOOD and GLAUERT 1960 a, b).

While the developmental character of the cycle described above in terms of the definitions given previously is quite apparent, it is illustrated also by the discovery of a proteinaceous factor produced by one strain and inducing conidia formation in another (SZABO et al. 1963).

For a recent summary of work on *S. coelicolor* the reader is referred to HOPWOOD (1967).

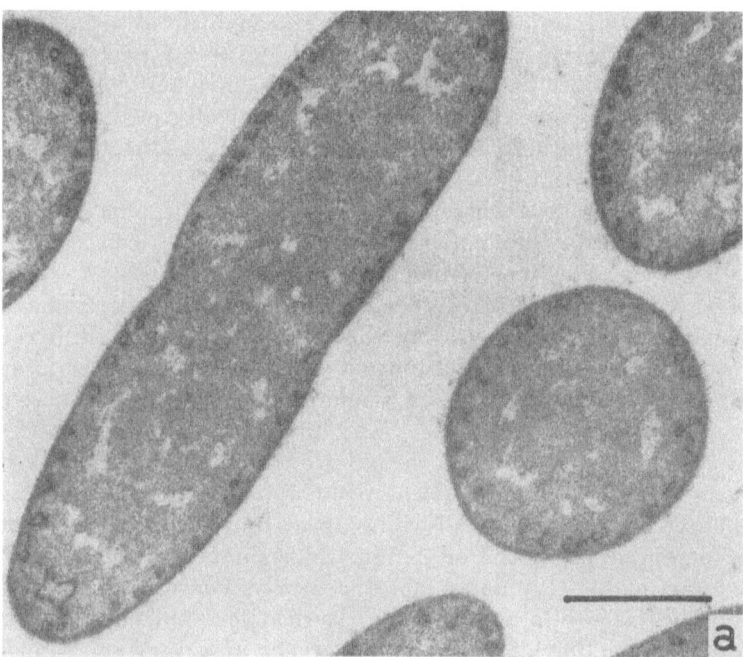

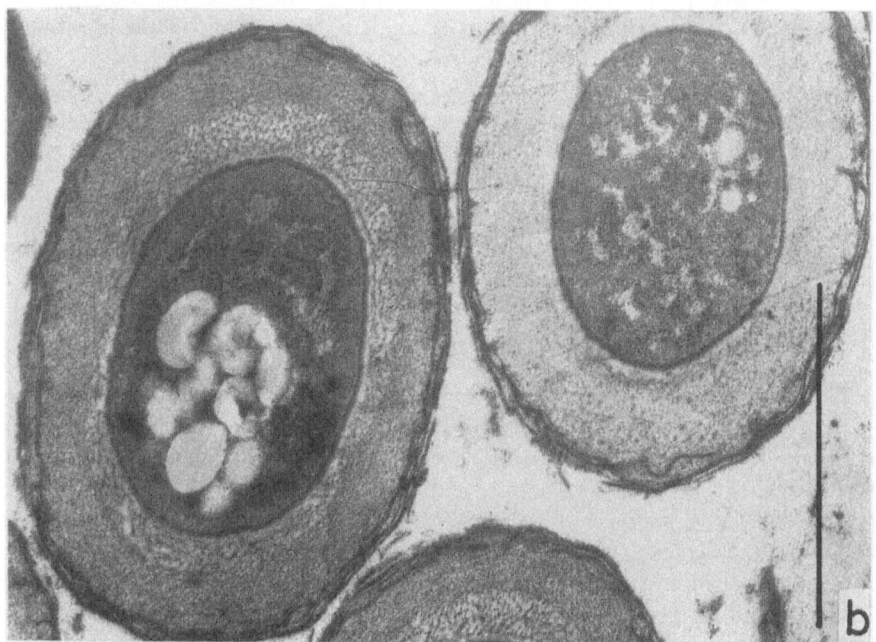

Fig. 30. *Azotobacter vinelandii*, RK-fixation. "Open" or expanded state of nuclear material. — *a* Vegetative cells. — *b* Mature cysts. — Scale markers represent 1.0 μ. Reprinted from Wyss et al. (1961).

5. Resting Stages in Nonsporing Eubacteria

In the past, bacterial cytologists have been little aware of the peculiarities of slow growing, starving, or resting nonsporing bacteria, and studies of the metabolic state of these forms until recently were very scarce. This is in marked contrast with the fact that practically all bacteria found in the soil, aqueous, and human environments are in some state of growth limitation or at least metabolizing at a reduced rate.

Extensive studies on bacteria from ageing cultures have been carried out by BISSET and his group (BISSET 1948 a, d, e, 1949, 1950, 1952 d, e, BISSET et al. 1951). Changes in morphology and nuclear patterns were similar in a wide variety of species (*Escherichia, Aerobacter, Salmonella, Lactobacillus, Rhizobium*). The cells turned from normal rod shape into resting cells which were smaller and shorter, even ovoid or spherical. If several nucleoids were present in the rod-shaped bacterium, they tended to be reduced to one by cell division, or, according to BISSET, may fuse into a single axial filament and to convert into a single round nuclear body. In some cases, the compound nuclear body appeared to separate again into two separate nucleoids, one of which in certain strains disappeared (suggesting degeneration of one of the nucleoids similar to events during endospore formation). As pointed out repeatedly, considerable care must be exercised in interpreting such results. Methodological considerations should be applied indicating that the disappearance of a very small nuclear element in stained preparations is not sufficient proof of its degradation (see also p. 84). For a critical evaluation of nuclear fusion the reader is referred to the previous sections of this review.

Azotobacter species develop a somewhat more elaborate type of resting stage referred to as c y s t. This structure essentially is a resting cell surrounded by two voluminous coats. Cyst wall and membrane are identical with those of the vegetative cells. The nuclear body appears in a more or less "open" form (Fig. 30) (WYSS et al. 1961, for light-microscopic observations see POCHON et al. 1948 and FLOETHMANN 1954). The cysts of *Azotobacter* may be related to the microcysts of myxobacteria (see below).

6. Nuclear Patterns during Morphogenesis of Myxobacteria

The formation of fruiting bodies in myxobacteria is accompanied by the conversion of cells into m i c r o c y s t s that resemble resting stages of eubacteria in that they are more or less spherical. Some resemblances exist to the cysts of *Azotobacter,* although the appearance of the extracellular layers is somewhat different (for details see DWORKIN and VOELZ 1962, VOELZ and DWORKIN 1962, VOELZ 1966). Microcyst formation can be induced experimentally by exposure of vegetative cells to 0.5 M glycerol (DWORKIN and SADLER 1966). After microcyst formation is induced in this manner, DNA synthesis amounts to approximately 20 per cent of the quantity already present (ROSENBERG et al. 1967).

Light-microscopic observations of nuclear changes are in general agreement with each other. The nuclear material is converted into a single central mass identical with the axial filament described above. Microcysts are believed to contain one or two nucleoids each (BADIAN 1930, BEEBE 1941, KLIENEBERGER-NOBEL 1947 b, GRACE 1951, FRUCHTER 1960, BAUER 1962).

On the electron-microscopic level of observation, some confusion was created by the interpretation of opaque inclusions in the nuclear areas as the nuclear bodies proper (IMZHENETSKIJ and ALFEROV 1962, for criticism see KÜHLWEIN and ROSSNER 1963). The opaque inclusions most probably are polyphosphate granules (VOELZ et al. 1966) (Fig. 69, p. 128).

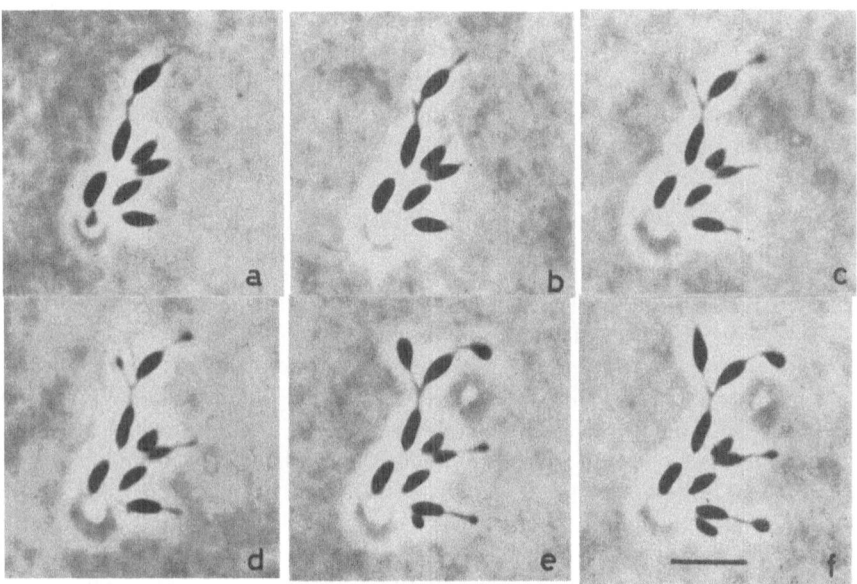

Fig. 31. *Rhodomicrobium vannielii*. Series of phase contrast micrographs showing the development of buds and branching. — *a* 0 min. — *b* 30 min. — *c* 1 hr. — *d* 1 hr 30 min. — *e* 5 hr. — *f* 7 hr 30 min. — Scale marker represents 5 μ. Reprinted from MURRAY and DOUGLAS (1950).

7. Stalked Bacteria

The state of the genera *Caulobacter* and *Asticcocaulis* including their cytology has been extensively reviewed by POINDEXTER (1964) and POINDEXTER and COHEN-BAZIRE (1964). The stalk is an extension of the cell proper and contains cytoplasmic material. It does not contain DNA. Structure and arrangement of the nucleoids follow the usual pattern.

8. Hyphomicrobium and Rhodomicrobium

The genera *Hyphomicrobium* and *Rhodomicrobium* resemble those mentioned in the previous section except for their mode of division. Daughter cells arise from the ends of the stalk-like extensions which here are referred to as hyphae. MURRAY and DOUGLAS (1950) report that small buds are formed at the free ends of the hyphae which at first are devoid of nuclear material, but formation of the bud apparently requires prior division of the nucleoid in the mother cell at the opposite end of the respective hyphal section. After some time, a minute nuclear element becomes apparent in the growing bud. It increases in size until it attains the size of a normal nucleoid. Its buildup apparently proceeds at the expense of one of the nucleoids in the mother cell. It must be assumed that the genophor of this nucleoid is reeled off and moves through the core of the hypha into the growing daughter cell where it again assumes the appearance of a distinct nuclear body. As a rule, one cell gives rise to one daughter cell at a time, but exceptions do occur. In *Rhodomicrobium*, completed daughter cells are sealed off by a septum which is formed in the hypha. Occasionally the hyphae form branches. In this case one of the cells undergoes another cycle of nuclear division and furnishes a nucleoid for the new cell which is formed at the tip of the new branch (Fig. 31).

Observations by Zavarzin (1960 and review, 1961) and by Hirsch and Conti (1964) indicate a more complicated cycle involving the formation of star-like aggregates.

H. Miscellaneous Bacteria

This chapter deals with a variety of forms not referred to in earlier sections of this review. In fact, the nuclear pattern in bacteria can well be described in general terms and undergoes relatively little modification even in widely different tribes. In this way, nucleoids in *Rhodospirillum* closely resemble those in other Gram-negative bacteria (Hickman and Frenkel 1959, Cohen-Bazire and Kunisawa 1963, Fuhs, unpublished results).

The nuclear bodies of the large sulfur bacteria are branched (*Chromatium okeni*: Bütschli 1890, 1896, Delaporte 1939/40, Kran et al. 1963) or form a peripheral band (*Thiovolum maius*: Rouillier and Fauré-Fremiet 1957, Fauré-Fremiet and Rouillier 1958, de Boer et al. 1961). Delaporte (1939/40) gives an excellent review on these and a variety of other unusual forms and mentions resemblances between nuclear configurations found in the larger bacteria with those encountered in the Cyanophyceae.

Transversely oriented horseshoe-shaped nuclear bodies were observed in *Caryophanon latum* (Peshkoff 1946, Pringsheim and Robinow 1946, Bisset 1953 a, Tuffery 1955). The sporeforming, flagellated *Oscillospira guilliermondi* shows a similar pattern (Petit 1927, Delaporte 1934, Tuffery 1954 a, b).

The nuclear body of *Vitreoscilla* (a gliding bacterium of uncertain position) occurs in the "open" or the condensed form depending on the solute content of the medium (Costerton et al. 1961).

The giant bacilli and sporeforming spirilla described by Delaporte (1935, 1936 a, 1964 a, b) were mentioned in the introduction. These forms display an extended axial network of nuclear material, a part of which is incorporated into the endospore while the rest is doomed to perish. This character indicates that these organisms are polyenergidic (see also p. 138 and Figs. 17–19).

The unusual form, *Metabacterium polyspora* forms numerous spores per cell. Apparently all of the nuclear material enters the spores (Robinow 1957). This situation resembles that found in the endospore-forming Cyanophyceae.

Spirochetes have received little attention so far. The small members of the group undoubtedly are extremely difficult objects for caryological studies. In one instance, several nucleoids were observed which were quite regularly spaced; fragments of the organism containing one nucleoid were considered viable (SCHLOSSBERGER et al. 1950). Recently RITCHIE and ROBINSON (1967) in electron-microscopic studies found an axial nuclear body in a form resembling *Borrelia*.

J. Modified Bacteria and Bacteria-Related Groups

1. Large Bodies

Large bodies are pathological forms arising from inhibition of the synthesis of the mucopeptide by penicillin or other agents such as ethanol (10 per cent), lithium or cesium salts. Nuclear replication and division may proceed without inhibition which results in the formation of large multinucleated bodies of irregular shape. For details and illustrations the reader is referred to TULASNE et al. (1948 a), PRÉVOT and REYMOND (1948), PONTIERI (1956), ZAPF (1957), METZ (1958) and ANGELOV et al. (1963).

2. L Forms

L forms are named after Lister (see KLIENEBERGER-NOBEL 1951, TULASNE 1953, and DIENES 1963 for review). They are cellular entities derived from bacteria but different in that the cell wall is missing or reduced to a template moiety (LEDERBERG 1956, LEDERBERG and ST. CLAIR 1958, LANDMAN and GINOZA 1961 a, b). L forms may arise as more or less stable clones from bacteria under the action of penicillin with large bodies as an intermediate stage. The term "stable protoplast (spheroplast) form" is often used synonymously with "L form".

Nuclear material in L forms was discovered by LIEBERMEISTER (1960), PESHKOFF and MOTUZOVA (1963), MOTUZOVA (1963), ERSHOV (1963) and WEIBULL (1965). COUSSONS and COLE (1967) found that the minimal size of a colony-forming L body has a volume that closely corresponds to that of a nuclear body of the same bacterial species carrying the minimal DNA complement. The worthwhile attempt to obtain high-resolution micrographs of complete series of ultrathin sections of these simple forms has not been undertaken.

3. Mycoplasmataceae

This group comprises a number of parasitic forms also known as pleuropneumonia-like organisms (PPLO). They resemble bacterial L forms, but their relationship with the latter is doubtful. Speculations by VAN ITERSON and others that these forms are modified pathogenic bacteria of some well-known type were discouraged by the finding that the base sequences in the DNA of M y c o p l a s m a species and the suspected bacterial relatives were entirely different (McGEE et al. 1965, ROGUL et al. 1965, see p. 7 for outline of method). DNA accounts for 1.5 to 4 per cent of the dry weight of these small forms which probably is less than would be required for a bacterial genome (SMITH 1964). The DNA appears to be organized in the same manner as in bacteria. EDWARDS and FOGH (1960) found a more open form of the nucleoid, micrographs by VAN ITERSON and ROBINOW (1961) show the rounded, compact type (Fig. 32).

4. Rickettsiae

Rickettsiae contain DNA which is localized in a central structure resembling a bacterial nuclear body in every respect, although a clear picture is sometimes difficult to obtain with the light microscope (RIS and FOX 1949, WYATT and COHEN 1952). Enzymatic studies confirm the existence of a DNA containing nucleoid and a RNA-containing cytoplasm (WIGAND and PETERS 1954 b, TAMURA 1959). Thin sections reveal a nucleoid of the bacterial type (STOKER et al. 1956).

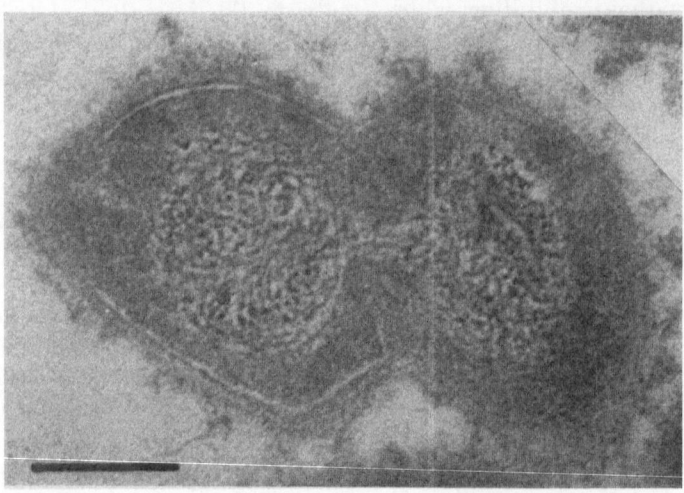

Fig. 32. *Mycoplasma sp.*, ultrathin section of dividing organism, showing bacterial type of organization of nuclear material. Fibrous state of DNA plasm was produced by reducing the time of exposure to Ryter and Kellenberger's fixative. Scale marker 0.1 μ. Reprinted from VAN ITERSON and ROBINOW (1961).

K. Nuclear Morphology as Seen in the Electron Microscope (Unsectioned Bacteria)

Early efforts to explore nuclear detail in unsectioned bacteria with the aid of the electron microscope gave unsatisfactory results and proved that this technique was not superior to light-microscopic investigation. In many instances, bacteria appear opaque without revealing any nuclear structures. In other instances, the outlines of the nuclear bodies become visible, the nuclear areas standing out as more transparent areas (ROBINOW and COSSLETT 1948, HILLIER et al. 1949, MUDD and SMITH 1950, KELLENBERGER 1951/52, 1952 b, KNAYSI et al. 1951, DE et al. 1953). Under conditions of nitrogen-limited growth, the cytoplasm becomes more transparent, and the nuclear bodies appear as dark areas (KNAYSI and BAKER 1947). Under such circumstances, however, the tendency of the cell to accumulate polyphosphates in granular form is greatly increased, and the granules are easily confused with nucleoids (see e.g. KNÖSEL 1963). Upon closer examination, distinctive characters can be found which consist in the sharper outlines of the granules as compared with nucleoids, and in the tendency of the granules to evaporate at higher electron densities.

In shadowed preparations the nuclear sites appear as depressions in the bacterial surface. This is a dehydration artifact caused by the lower solids

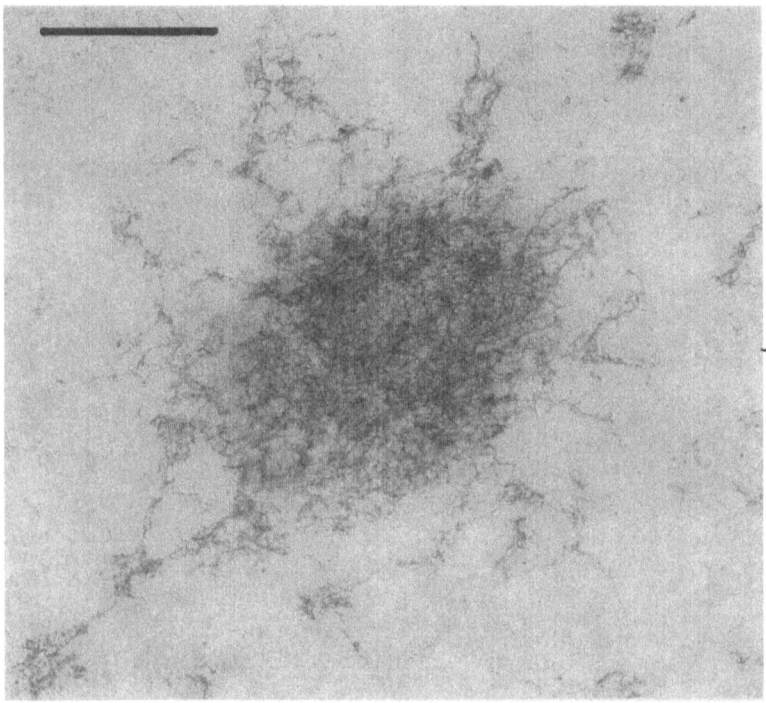

Fig. 33. Ultrathin section of nuclear body of *Bacillus megaterium* isolated with technique described by FITZ-JAMES (1964). Original micrograph by Dr. P. C. FITZ-JAMES.

content of the nuclear bodies (WINKLER and KNOCH 1951, WINKLER et al. 1951). Negative staining does not reveal any nuclear detail despite recent claims to the contrary by MEYER (1965).

After removal of ribonucleic acid and protein components by ribonuclease-pepsin, the nucleoids stand out clearly as opaque areas in a transparent cytoplasm. In shadowed preparations, they appear as elevated areas within the partly empty, collapsed cell body (PETERS and WIGAND 1953, EHRHART and STEIGLER 1954). The positive identification of the nucleoids in such preparations is accomplished by deoxyribonuclease treatment or treatment with certain impure (not nuclease-free) trypsin preparations. Polyphosphate granules, contrary to nuclear bodies, resist such treatment (see KNÖSEL 1963 for illustrations).

L. The Isolation of Nuclear Bodies

After a first but not very successful attempt by MARSHAK (1951 a, b), FITZ-JAMES (1958, 1964 b, also SPIEGELMAN et al. 1958), succeeded in isolating nuclear bodies of *B. megaterium* by lipase treatment of protoplasts under controlled conditions and under critical microscopic examination during all phases of the experiment. In ultrathin sections the nuclear bodies appeared as a tangled but coherent aggregate of DNA fibres (FITZ-JAMES 1964 b) (Fig. 33). Chemical analysis revealed also RNA and certain protein constituents which were believed to be located in the microscopically discernible "core" (see also p. 127).

DELAMATER (1959) and ECHLIN and DELAMATER (1962) found that treatment of *B. megaterium* cells with glycocholate resulted in extensive leakage of various cell constituents into the surrounding medium while the DNA remained inside. After isolation, the DNA was found to be associated with basic protein components. This association, however, may represent an artifact (see p. 52).

III. Nucleoid Fine Structure as Seen in Ultrathin Sections

A. Introduction

Since electron microscopic observation of whole bacteria failed to reveal nuclear fine structure, ultrathin sectioning appeared to be the most promising technique of investigation. But the results obtained during the first decade of its application differed widely depending on the procedure used.

At the present time, when establishment of a consolidated picture seems possible, two erroneous assumptions implicitly made by many authors appear to be responsible for the divergence of interpretation of the structural patterns obtained. The first of these assumptions is that valid criteria for the interpretation of ultrastructure in the macromolecular range could be easily derived from light microscopic experience, the other, that a single procedure could be developed that at the same time rigidly preserved and rendered visible the *in vivo* arrangement of the DNA.

The difficulty of establishing valid criteria for the interpretation of ultrastructure is illustrated in the following examples taken from actual discussions in the field of bacterial caryology.

The "Order-or-Disorder" Argument. In light microscopy as well as electron microscopic cytology at moderate magnifications, apparent order is commonly taken as evidence of preservation of cytological detail, whereas distortion, disruption, and disintegration of cell structure are safe indicators of improper techniques. With regard to nucleoids, GIESBRECHT (1959) claims that only certain more regular arrangements of DNA material are artifact-free representations of nuclear fine structure, while apparent disorder indicates improper preservation of detail. SCHREIL (1964), on the other hand, concludes from *in vitro* experiments that certain fixatives artificially induce parallelism of fibres in a solution of DNA. Therefore, a random arrangement of fibres within the nuclear body might be considered closer to natural.

Evidently neither of these conflicting statements can be valid as a generalization.

The Coagulation-Disintegration Dilemma. Coagulation effects are a major menace in all investigations involving fixation and dehydration of cell material, and both types of treatment are necessary in the preparation of bacteria for ultrathin sectioning. RYTER and KELLENBERGER (1958) assume that a structureless appearance of the DNA-plasm most closely resembles the living state, since coagulation as an event that may distort the true picture certainly can be ruled out. On the other hand, destruction of any tertiary structure, dissociation from DNA of bound protein, and the partial or complete depolymerization or denaturation of DNA also would result in a more homogeneous appearance of the DNA-plasm. Obviously the latter possibility must be ruled out before the homogeneous aspect can be accepted as a true representation of the DNA-plasm.

Optimum Preservation vs. Interpretability. This point is best illustrated by an example: WOHLFARTH-BOTTERMANN (1961) in studies on the cytoplasm of amoebae describes the cytoplasmic "background" structure as a randomly arranged granular-fibrillar network. While one may hesitate to consider this finding in itself highly significant, a quite different pattern consisting of parallel fibres was observed in the ectoplasm of other rhizopods, in particular in such areas as are known to exhibit protoplasmic streaming. Whereas both the irregular and the ordered pattern may be due to coagulation on a reduced scale, the d i f f e r e n c e is significant and undoubtedly reflects a difference in molecular arrangement. WOHLFARTH-BOTTERMANN in dealing with this situation introduces terms such as "equivalent structure" and "equivalent value (of a structure as related to the unperturbated state of the cytoplasm)". It will be shown below that this concept is quite helpful in the interpretation of nucleoid ultrastructure.

Even in the wake of a period when preservation of mitochondrial detail was considered evidence for the successful demonstration of the cell in general, the literature is by no means free from arbitrary interpretation of effects of fixation and related preparatory measures. Any statement that a method "was found particularly suitable for nuclear structures (or membranes)" should be given very careful examination. Nor does preservation of structure in the medium range of resolution provide any proof that the macromolecular pattern is unaltered.

B. Types of Preservation of Nucleoid Fine Structure

1. Artificially Condensed States

If bacteria are fixed in buffered osmium tetroxide solutions, subsequently dehydrated in an organic solvent such as ethanol or acetone, and embedded in a suitable resin (methacrylates, Vestopal, Epon, etc.) according to procedures recommended and successfully used for sectioning of animal tissues (PALADE 1952), the nucleoids appear as electron-transparent zones which contain masses and strands of opaque material (CHAPMAN and HILLIER 1953, BIRCH-ANDERSEN et al. 1953, BRADFIELD 1954, TOMLIN and MAY 1955, KELLEN-

BERGER and RYTER 1955, PIEKARSKI and GIESBRECHT 1955, 1956, BRIEGER and
GLAUERT 1956, GLAUERT and BRIEGER 1956) (Fig. 34). Micrographs of this
type were often considered to show a chromosome-like core surrounded by
a layer of unidentified material (MAALØE and BIRCH-ANDERSEN 1956, GIESBRECHT
1958, GIESBRECHT and PIEKARSKI 1958, PIEKARSKI and PONTIERI 1956, HIGASHI
1959). Digestion with hydrolyzing acids and with deoxyribonuclease
showed that the opaque structures are DNA (KAWATA 1958, FUKUSHI 1959).

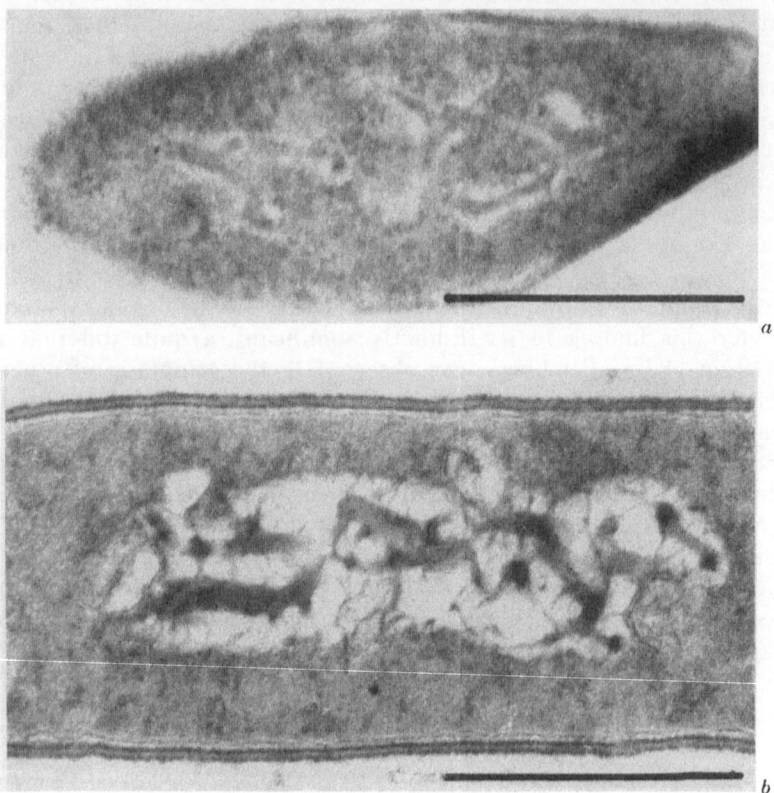

Fig. 34. *Bacillus subtilis*, artificially condensed state of DNA plasm. Cells fixed in osmium-dichromate without
subsequent treatment with uranyl acetate solutions. *a* "open" or expanded state, *b* compact state of nucleoid.
Scale markers: 0.5 μ.

By comparative light and electron microscopy evidence was presented
that the nucleoid as defined by DNA-specific reactions in the light microscope corresponded in size and shape with the entire electron-transparent
vacuole rather than with the opaque strands and masses (BIRCH-ANDERSEN
1955, KUSHNAREV 1959). This favored the view that the opaque masses could
be the shrunken nuclear bodies. Failure of the fixative to react with DNA
(BAHR 1954, DE LOZÉ and LENORMANT 1959) might result in insufficient protection of the DNA-plasm against massive coagulation during dehydration
and embedding in plastic. These considerations led to the use of agents
which were known to precipitate nucleic acids and therefore could be considered fixatives for DNA. If the fixative was a heavy metal, its location

could be detected in the electron microscope. This favored agents such as lanthanum and uranyl salts as precipitants and electron stains for DNA. Both were introduced into bacterial fine structure research by KELLENBERGER (KELLENBERGER and RYTER 1955, 1956).

Later experiments showed that KELLENBERGER's assumptions were correct, in that the opaque masses were shrunken nuclear bodies. The excessive shrinking occurred during dehydration and was reversible. By partial dehydration and application of uranyl solutions of high ionic activity intermediate states could be produced. In this manner, it was shown that the moderately shrunken nuclear body had the shape of the "nuclear vacuole" (FUHS 1964) (Fig. 35).

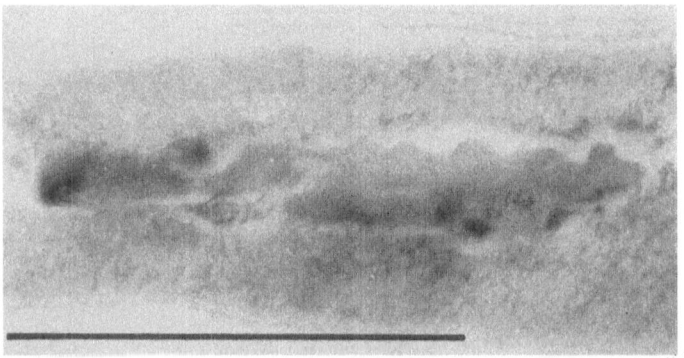

Fig. 35 a. *Bacillus subtilis*, fixed in osmium-dichromate mixture and partly dehydrated in 70% ethanol before treatment with solution of 2 percent uranyl nitrate in 70% ethanol. Partly shrunken contents of nuclear area reflect the shape of the "nuclear vacuole", shrinking of nucleoid with respect to overall shrinking of cell is less severe than in cell shown in Fig. 34 a and 34 b. Scale marker: 0.5 μ. — Reprinted from FUHS (1964).

2. Fibrous State

The use of uranyl or lanthanum salts for the stabilization of nucleic acids profoundly affects nucleoid fine structure (KELLENBERGER and RYTER 1956, RYTER and KELLENBERGER 1957, 1958, KELLENBERGER et al. 1958 a, b, GIESBRECHT 1958, KRAN and SCHLOTE 1959). If used following osmium or osmium-dichromate fixation, no major empty spaces are seen. The nucleoid region is filled with fibrous material. The diameter of individual fibres varies from approximately 50 to 150 Å. The fibres are degraded by deoxyribonuclease and therefore must essentially consist of DNA (LEE 1960). This situation is illustrated in Figs. 36 and 38 as well as in other illustrations presented in the later sections.

While some authors consider the fibrous state to be caused by a moderate aggregation of DNA material (RYTER and KELLENBERGER 1957, 1958, FUHS 1964, 1965 d), others consider uranyl ions to exert a harmful effect on nucleoid fine structure (KRAN und SCHLOTE 1959). GIESBRECHT (1959 and later publications) in a rather arbitrary procedure selects certain patterns as reflecting the true arrangement of DNA material while disregarding others as allegedly reflecting the destructive action on nucleoid structure of uranyl ions.

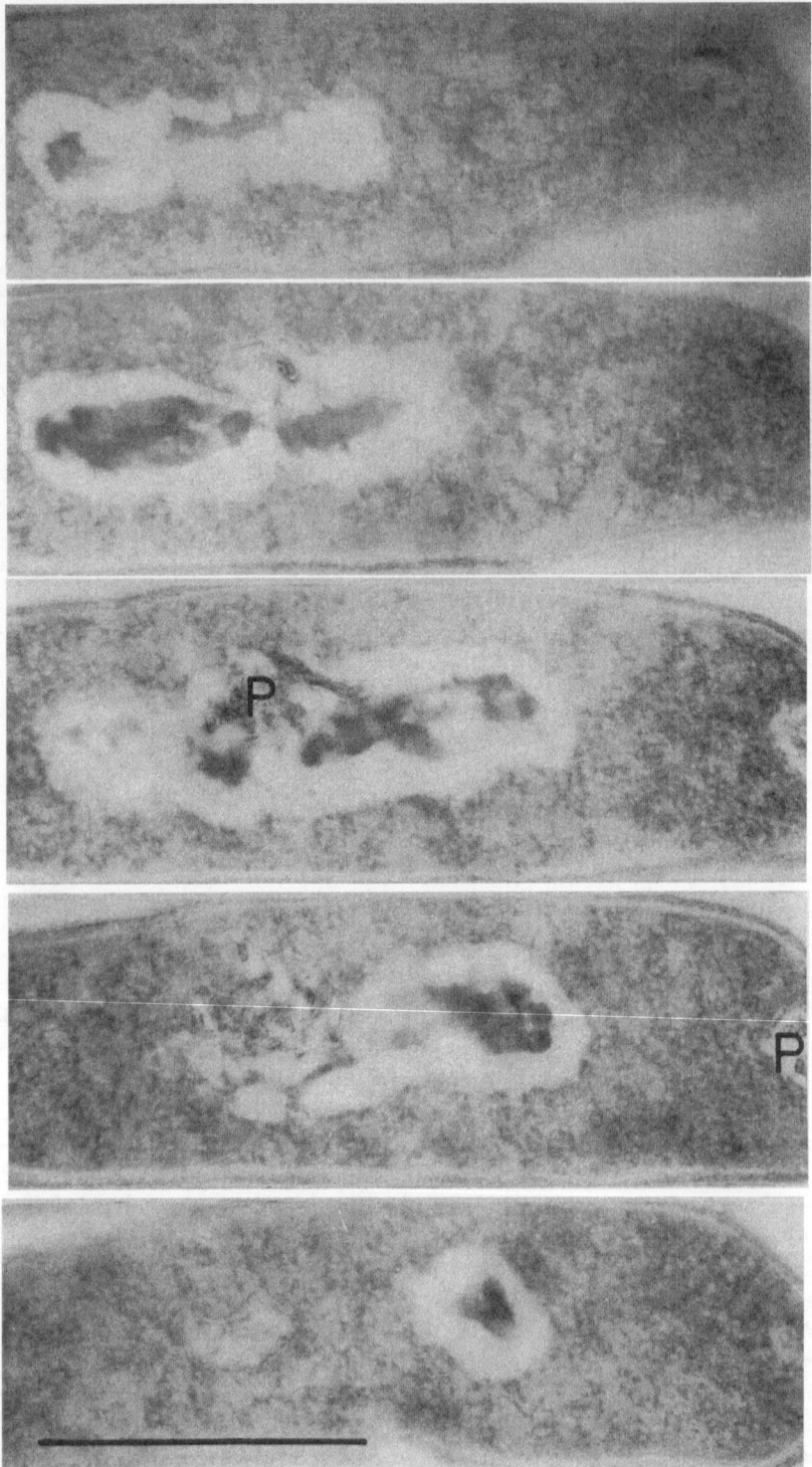

Fig. 35 *b*. *Bacillus subtilis*, same technique as in Fig. 35 *a*. Serially sectioned cell showing two strands of nuclear material which appear to emerge from the central plasmalemmosome (*P*). They extend towards the poles and turn back towards the center of the cell where they may or may not be attached to the plasmalemmosome. See Figs. 50 and 52 for similar nuclear configurations. The polar plasmalemmosomes (only one shown) were without apparent connection to the nucleoids. The strands of nuclear material shown with this technique more likely correspond to the bundle-like formations of parallel DNA fibres than the severely disrupted structures shown in Figs. 34 *a* and 34 *b*. — Scale marker represents 0.5 µ.

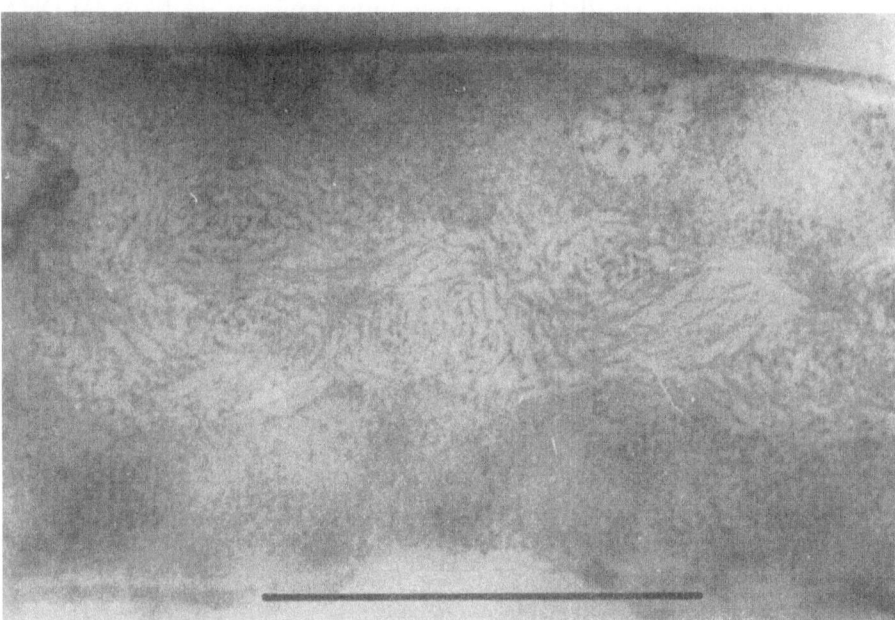

Fig. 36. *Bacillus subtilis*, fibrous state of DNA plasm. Osmium-dichromate fixation, uranyl acetate applied in aqueous solution, dehydration in acetone, Vestopal embedding. Scale marker represents 0.5 μ.

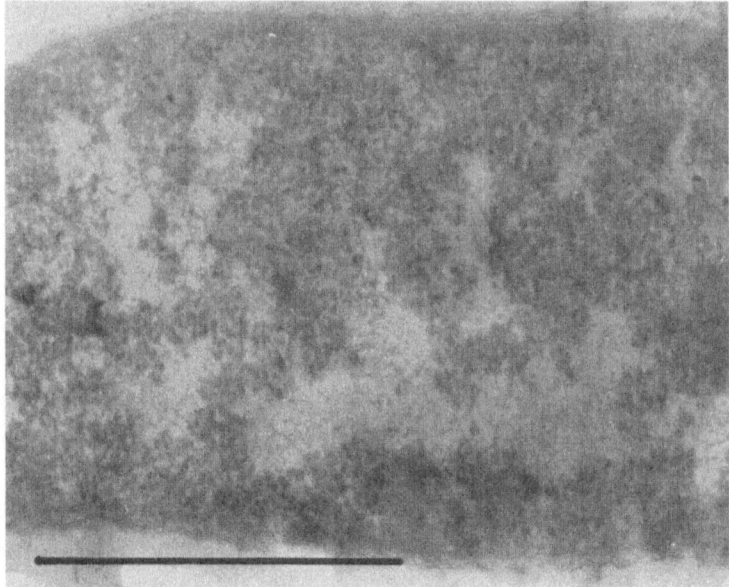

Fig. 37. *Bacillus subtilis*, homogeneous state of DNA plasm. RK (Ryter-Kellenberger) fixation, uranyl acetate, Vestopal. Scale marker represents 0.5 μ.

3. Homogeneous State

A more or less perfectly homogeneous appearance of the bacterial DNA-plasm is obtained by using the fixation technique by Ryter and Kellen-berger (1958). This technique is discussed in considerable detail below and

Fig. 38. *Escherichia coli*, fibrous state of DNA plasm. Osmium-dichromate fixation, uranyl acetate, Vestopal. — Scale marker represents 0.5 μ.

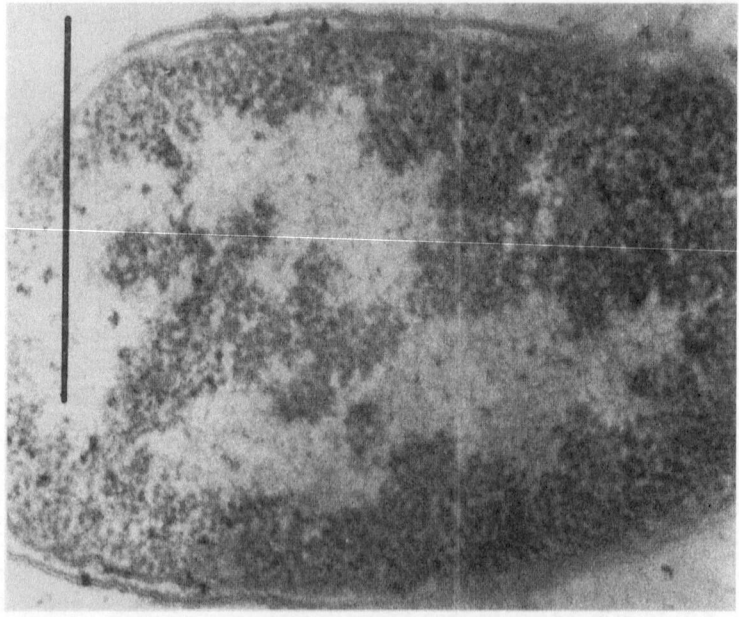

Fig. 39. *Escherichia coli*, homogeneous state of DNA plasm. RK (Ryter-Kellenberger) fixation, uranyl acetate, Vestopal. To enhance contrast, sections were left floating on saturated uranyl acetate solution for several hours. The black deposits are an uncalled-for side effect of this treatment. Scale marker: 0.5 μ.

is commonly referred to as RK-technique and includes modified osmium fixation followed by uranyl treatment. The homogeneous state is shown in Figs. 37 and 39. As can be seen in the illustrations, the new method also gives the best overall preservation of cellular detail. A possible explanation for this is given in the next section. According to KELLENBERGER, the RK-technique avoids coagulation artifacts to a considerable degree. Arguments that the resulting apparent lack in visible organization was due to degradation of a hypothetical tertiary structure or the secondary structure or extensive depolymerization of native DNA (KRAN and SCHLOTE 1959, GIESBRECHT 1959) was disproved (FUHS 1964, 1965 a).

C. Effects on Nucleoid Structure of Fixation and Embedding Techniques

This section is an attempt to describe in greater detail the various and sometimes profound effects that preparatory measures are known to have on nucleoid fine structure.

1. Osmium Fixation

Osmium tetroxide alone or in combination with potassium dichromate is used in a variety of mixtures which are essentially derived from formulas given by PALADE (1952) and DALTON (1955) (see FUHS 1964 for details). The use of dichromate is of minor importance except that its use permits the mixture to be compounded from stable stock solutions. Addition of dichromate to a balanced fixative of the Palade type also causes changes in ionic strength and buffer capacity. The useful pH range is 6.0 to 7.2.

If solutions of DNA and osmium tetroxide (with or without dichromate) are mixed in a test tube, no reaction becomes apparent. The DNA remains in solution, and osmium stays in its oxidized form (BAHR 1954). The infrared spectrum of the DNA is not changed (DE LOZÉ and LENORMANT 1959). The viscosity of the mixture does not change in any measurable degree indicating that the DNA is neither depolymerized nor denatured (FUHS 1965 a).

While deoxyribonucleohistones may be susceptible to osmium tetroxide fixation due to a reaction with the histone (BAHR 1954, DE LOZÉ and LENORMANT 1959, SCHREIL 1964), the polyamines that presumably are associated with bacterial DNA do not react with osmium tetroxide. It is evident from these observations that treatment with osmium tetroxide solutions in itself does not result in stabilization of the DNA-plasm.

Apparently such stabilization is achieved in the Ryter-Kellenberger (RK) fixation procedure. In this technique, the osmium tetroxide solution contains a buffer system, calcium ions, and amino acids in the form of tryptone (RYTER and KELLENBERGER 1958). This mixture can transform a solution of DNA into a gel-like precipitate (SCHREIL 1961, 1964). It has been suggested that amino acids and calcium ions mediate a reaction between DNA and osmium tetroxide.

Contrary to ordinary osmium fixation, RK-fixation results in the introduction and deposition of osmium in the DNA-plasm. This is apparent from a change in the mass density pattern of the bacterial cell as seen in the phase-contrast microscope (FUHS, unpublished results). This author also found that tryptone as a component in the RK fixative could be effectively substituted by either arginine, lysine, glutamic or aspartic acids. Each of these agents produced a more pronounced change in the mass density pattern of the cell than did tryptone. The homogeneous aspect of the DNA-plasm as obtained with the modified RK-fixative is apparent from Fig. 40.

The four amino acids mentioned in the preceding paragraph reduce osmium tetroxide as is evident from the production of a dark color or a dark precipitate in osmium tetroxide solutions (Bahr 1954). Only the basic amino acids, arginine and lysine, react with DNA, but the association is quite weak and calcium ions in the mixture may easily interfere with its

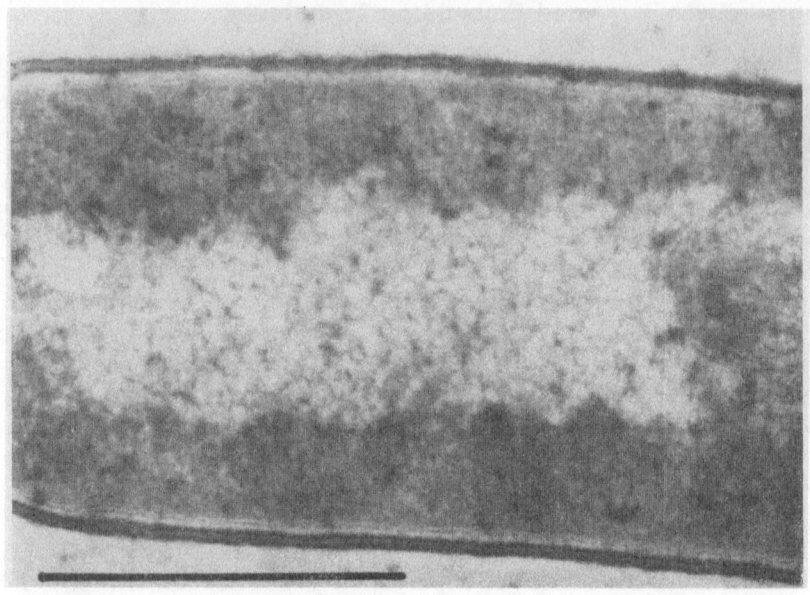

Fig. 40. *Bacillus subtilis*, homogeneous state of DNA plasm as obtained by a modification of the RK fixation consisting in substitution of tryptone by 0.5 per cent of each arginine and lysine. Scale marker: 0.5 μ.

formation (see p. 50 for references). On the other hand, calcium ions may mediate complex formation between DNA and glutamic and aspartic acids as suggested by Huxley and Zubay (1961). It appears from these considerations that the chemical basis of the RK-fixation is not clear.

Kellenberger (personal communication) points out that the RK-reaction invariably is accompanied by a blackening of the mixture containing the bacteria and the fixative. Failure of the mixture to turn dark invariably resulted in a failure to establish the homogeneous state of the DNA-plasm. According to Kellenberger, the onset of the reaction as evident from the darkening of the mixture may occur at different times even in replicate samples. In the present author's laboratory, the reaction was usually complete after two or three hours at room temperature. Kellenberger recommends fixation over night.

The conditions of the RK-fixation as originally published apply to *B. subtilis* as well as *E. coli* cells (Figs. 37 and 39) but not necessarily to other bacteria (e.g. spirochetes, Ritchie and Robinson 1967). In fact, the

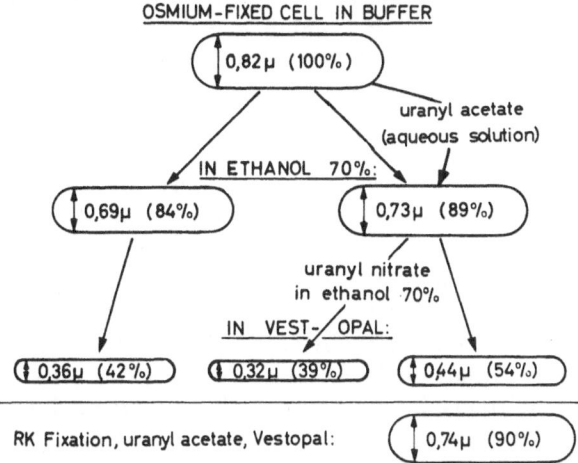

Fig. 41. Reduction in diameter during dehydration of bacterial cells fixed with the osmium-dichromate technique and treated with uranyl salt solutions prior to or after partial dehydration or according to the Ryter-Kellenberger procedure. After FUHS (1964), slightly modified.

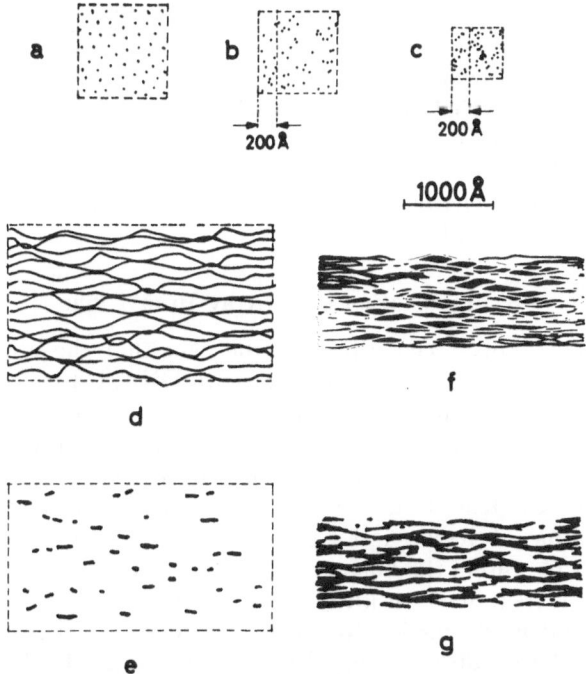

Fig. 42. Changes in the bacterial DNA plasm during preparation and resulting optical artifacts. — a DNA plasm *in vivo*, DNA fibres shown in cross-section, drawn to scale. — b Same as in a, after RK fixation, uranyl acetate treatment, dehydration and embedding in plastic. — c Same as in a, but after fixation with osmium-dichromate, uranyl acetate treatment, dehydration and embedding in plastic. — d Face view of 200 Å thick section cut from RK-fixed material as shown in b. — e Approximate appearance of section shown in Fig. d if electron-optical conditions (contrast and resolving power) do not permit the recognition of single 20 Å fibres. This pattern resembles the "homogeneous state" as seen in the electron micrographs. — f Face view of 200 Å thick section cut from osmium-dichromate-fixed material as shown in c. — g Appearance of section shown in Fig. f under electron-optical conditions described under e. Pattern resembles "fibrous state" as seen in the electron micrographs. — Reprinted from FUHS (1964) with minor modification.

conditions for the RK-reaction may be quite critical, and even a superficial survey of the literature shows a number of illustrations of bacteria which are said to have been subjected to the RK-fixation procedure, the nucleoids of which, however, appear in the fibrous rather than in the homogeneous state. Appearance of the fibrous state indicates failure to establish proper conditions for the RK-reaction.

Fuhs (1964) observed that RK-fixation markedly reduces the tendency of the bacterium to shrink during subsequent dehydration. In ultrathin sections, the diameter of bacteria fixed according to Ryter and Kellen-berger was 90 per cent of that of the living bacterium, while after ordinary osmium fixation and uranyl treatment the diameter was reduced to 54 per cent (see Fig. 41). This most obviously will contribute towards a more homogeneous appearance of the DNA-plasm as is evident from a considera-tion of optical artifacts in a fibrous pattern as related to the degree of shrinking (Fig. 42).

2. Fixation with Agents Other than Osmium Tetroxide

Formaldehyde resembles osmium tetroxide in not reacting with native DNA. This is due to the fact that the amino groups of the DNA bases in the native state are hydrogen-bonded and unavailable for reaction. Possible candidates for a reaction with the aldehyde are the polyamines, and such reaction may act towards a stabilization of nucleoid structure. *Glutar-aldehyde* can be expected to react similarly but with less impact on cellular structures in general due to its lower osmotic activity. The agent is being used successfully in the study of Cyanophyceae (see e.g. Echlin 1964). An example of glutaraldehyde-osmium tetroxide fixation in bacteria is given by Jordan and Grinyer (1965) (Fig. 43).

3. Stabilization of DNA by Uranyl Ions

As indicated above, recognition of the fact that the otherwise useful fixative, osmium tetroxide, failed to react with DNA, led to the introduction of uranyl acetate as an agent capable of precipitating (and therefore "fixing") DNA. After ordinary osmium fixation (according to Palade or Dalton), uranyl acetate application results in the appearance of the fibrous rather than the condensed state of the DNA-plasm. But also the modified osmium technique of Ryter and Kellenberger relies on the action of the uranyl salt for obtaining the homogeneous aspect of the DNA-plasm. Omis-sion of uranyl treatment results in the appearance of empty nuclear spaces and condensed states with very little contrast (Fuhs, unpublished results). This latter observation also indicates that the introduction of osmium into the nuclear areas during RK-fixation contributes relatively little towards the electron staining of the DNA.

While historically fixation is synonymous with precipitation or coagula-tion (denaturation) of living, especially proteinaceous matter, considera-tions of effects on ultrastructure require a somewhat refined terminology. As will be shown below, the DNA-plasm essentially consists of hydrated

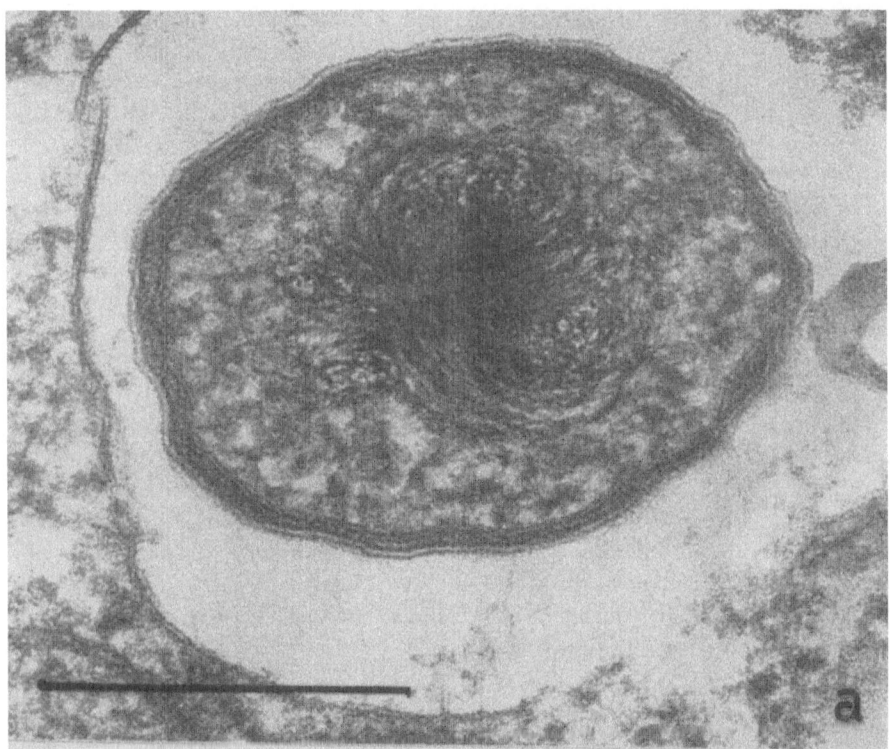

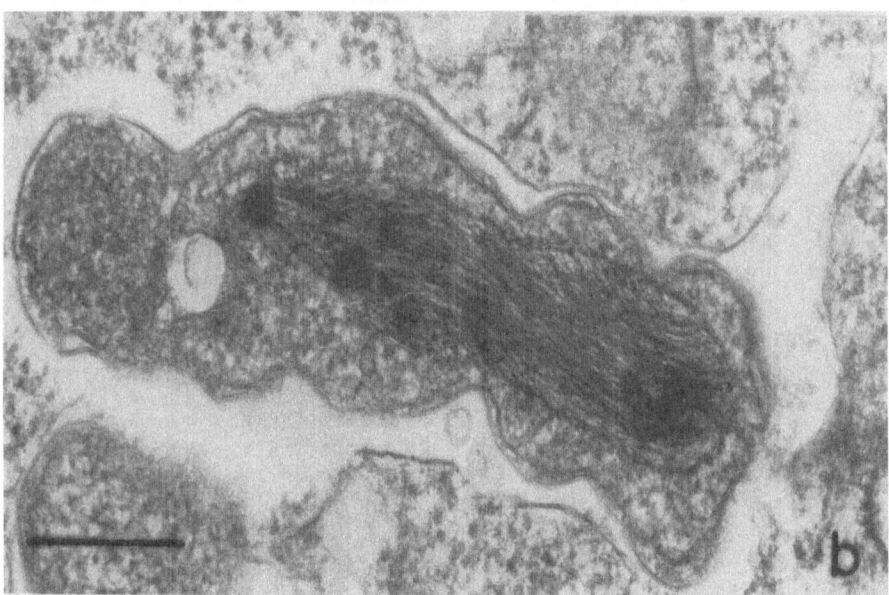

Fig. 43. Bacteroids within a lupine host cell. Excised material fixed with 5% glutaraldehyde in 0.1 M phosphate buffer (pH 7.3, 3 hours), washed over night in 0.2 M sucrose in buffer, postfixation in 1% buffered osmium tetroxide (2 hours), saturated uranyl acetate solution (2 hours), acetone. Embedded in mixture of Selectron resins. Fibrous state of DNA plasm. — a The nuclear material has been cut in cross section. — b Same, longitudinal section. — Scale markers indicate 0.25 μ. Reproduced from JORDAN and GRINYER (1965) by permission, National Research Council of Canada.

DNA fibres that are oriented roughly parallel to each other. Substitution of water by an organic solvent has a similar effect as is observed with preparations of DNA *in vitro:* large-scale side-by-side aggregation of DNA molecular fibres will result in the formation of a solid, dehydrated structure. Theoretically the ideal way of stabilizing the DNA-plasm would be the introduction of an agent that would establish numerous bridges or cross-links between neighboring fibres and in so doing would render the DNA-plasm sufficiently stable to withstand removal of bound water without collapse of structure. While the operation of such mechanism during RK-fixation is still a matter of speculation, uranyl ions certainly do not act that way. Uranyl ions react with DNA phosphate groups to form a strong complex the association constant of which is in the range of 10^6 (ZOBEL and BEER 1961, FUHS 1965 c). Uranyl complex formation effectively interferes with the binding of water and drastically reduces the free surface charge of the DNA polyanion. Elimination of surface charge is equivalent to the elimination of repelling forces between neighboring DNA fibres, and some small-scale aggregation may occur even in the presence of water. (This effect corresponds to the formation of a precipitate in *in vitro* experiments and is not dependent on cross-linking.) The resulting precipitate is essentially inert with respect to its interstitial, almost unbound, water which may be exchanged against a less polar solvent without further collapse of structure.

Fuhs (1964, 1965 c) found that the stabilization of the bacterial DNA-plasm requires neutralization of approximately one half of the DNA phosphate groups by uranyl ions. The external concentration of uranyl ions (uranyl ion activity) is 5×10^{-4} molar. This value is higher than would be expected from the high association constant. The discrepancy is explained by the possible presence of competing ligands within the bacterial cell (FUHS 1965 c) (see p. 53).

Uranyl ion activities of 5×10^{-4} molar and above can be obtained in aqueous solutions only. Solutions containing uranyl acetate in ethanol as frequently recommended for general cytological use are worthless for the preservation of bacterial nuclear structures (FUHS 1964). On the other hand, transfer of specimens after uranyl treatment into moderately concentrated ethanol or acetone is essential for the preservation of the uranyl-DNA complex. The low constant of dissociation of uranyl salts in such solvents discourages dissociation of the complex, while dissociation according to the law of mass action will occur in aqueous washings following uranyl treatment.

The pattern of aggregation as produced by uranyl action in a mass of randomly oriented DNA fibres is markedly affected by the direction and rapidity of penetration by the uranyl solution as well as by its concentration (SCHREIL 1961, 1964). Interpretation of such patterns therefore requires proof regarding the preservation of the natural arrangement of DNA fibres. Evidence that neither denaturation nor extensive depolymerization of DNA occurs during preparatory treatment would constitute an important part of such proof, since preservation of the genophor as a coherent two-stranded DNA structure would markedly reduce chances of major rearrangement of DNA material, and mechanisms of rearrangement requiring single or two-strand breakage could be ruled out as a possibility.

Preliminary evidence that the integrity of two-stranded DNA was preserved during osmium and uranyl treatments was obtained by viscosimetry of DNA solutions which were treated accordingly, and in a more definitive form in experiments on the experimental denaturation of intracellular DNA by heat (Fuhs 1965 a). In the viscosimetric experiments, the usual treatment with 0.25 per cent uranyl acetate did not result in a noticeable decrease of viscosity nor in a change in the non-Newtonian behavior of the DNA solution. If treatment was extended beyond the normal one or two hour period to 24 hours, a slight decrease in viscosity became apparent. If pH during treatment was lowered to a value below 3. e.g. by the use of uranyl nitrate rather than uranyl acetate, viscosity decreased quite rapidly. This shows that uranyl complex formation does not protect DNA from degradation at low pH values. The same holds for denaturation by heat. Exposure of the insoluble uranyl-DNA complex to temperatures above the

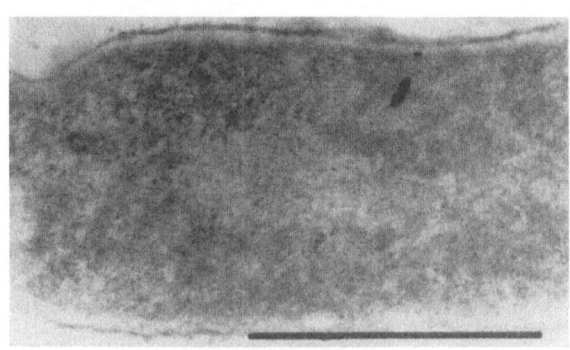

Fig. 44. *Bacillus subtilis*. Osmium-dichromate fixation followed by heating to 96° C in standard 0.15 M NaCl-0.015 M sodium citrate buffer for 40 minutes. Uranyl acetate, acetone, Vestopal. DNA denatured, fibrous state no longer apparent. Scale marker 0.5 μ.

melting point of the DNA secondary structure apparently results in strand separation and, upon cooling, in extensive cross-linking of strands. This is apparent from the insolubility of the heated precipitate in the presence of a strong chelating agent such as ethylene-diamine tetra-acetic acid. Such experiments show that low concentrations of uranyl ions do not interfere with the two-stranded conformation or with the degree of polymerization of DNA. This is contrary to the action of very concentrated solutions which have been used to extract DNA from vaccinia virus (Chatterjee and Sarkar 1962).

These findings were confirmed and extended in experiments on the denaturation of intracellular DNA of bacteria. In bacteria fixed with osmium tetroxide-dichromate mixtures and treated with uranyl acetate solutions, a marked structural change in the nuclear areas could be induced by exposure to pH values or temperatures that are known to cause denaturation of DNA *in vitro*. The effect consists in a loss of order in the nuclear pattern and the appearance of a new aspect characterized by a random or chaotic arrangement of the DNA fibres (Figs. 44, 45, and 46). Fig. 46 also indicates a loss of DNA material during extended treatment with a uranyl nitrate solution of pH 3.0. Loss of parallelism and appearance of a random arrangement was considered to indicate transition of the DNA from the two-

stranded helical to the single-stranded coiled state.—It was also found that
intracellular DNA required much more time for the transition than did
low molecular weight DNA *in vitro.* This was explained in terms of a
requirement for single-strand breaks in the coherent helix (the genophor)

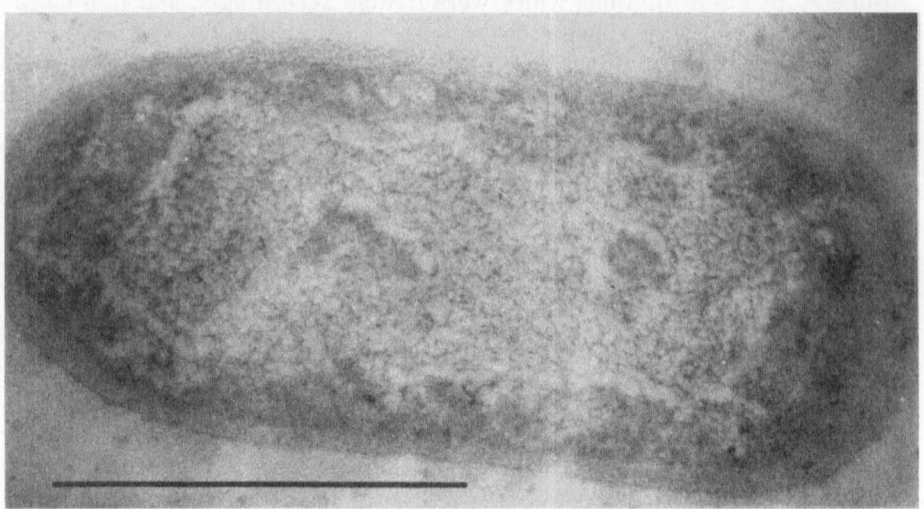

Fig. 45. *Escherichia coli.* Treatment of cells as described in preceding Figure. DNA denatured, fibrous state no
longer apparent (compare with Fig. 38). Scale marker: 0.5 μ.

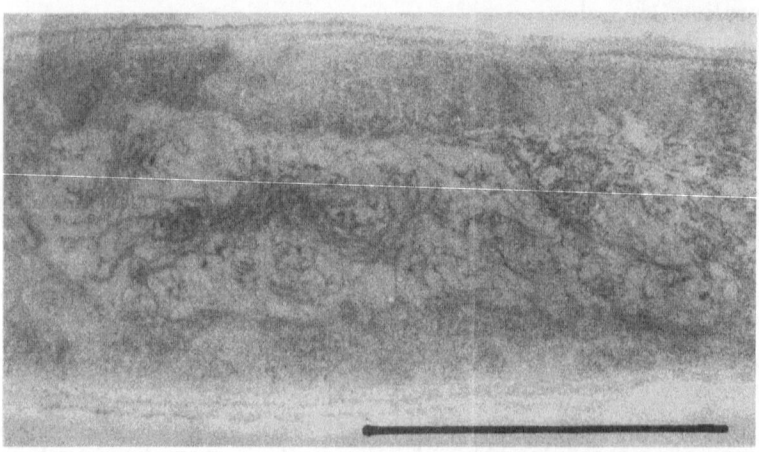

Fig. 46. *Bacillus subtilis.* DNA partly degraded (depolymerized and eliminated) by treatment with uranyl salt
at low pH (1% aqueous solution of uranyl nitrate, pH 3.0) for 2 hours. Fixation prior to uranyl treatment was in
osmium-dichromate, embedding in Vestopal. — Scale marker is 0.5 μ.

prior to unwinding. That this assumption was correct was shown in the
following manner: Formamide disrupts the hydrogen bonds in native DNA
without introducing chain breaks. It therefore can cause denaturation of
DNA *in vitro,* but was found not to induce transition from the ordered to
the chaotic appearance in intracellular DNA. Heating for several hours
of the cells to a temperature slightly below the melting point of the DNA

was equally ineffective in randomizing nucleoid fine structure. Such treatment can, however, be expected to cause numerous chain breaks in the
intracellular DNA (Doty et al. 1960, Eigner et al. 1961). If the heating step
was followed by treatment with formamide, or by raising the temperature
above the melting point of the DNA secondary structure, the time
required for a change in appearance of nucleoid fine structure no longer
exceeded that required for denaturation *in vitro* (less than 5 minutes). As
a conclusion one may safely assume that the molecular conformation of
intracellular bacterial DNA is not affected by uranyl ions, provided that
the concentration of uranyl salt does not exceed 2 per cent, the pH of the
uranyl solution is above 3.5, and the time of treatment does not exceed one
or two hours, longer exposure being permissible at higher pH values which
reduce the danger of slow depolymerization. These pH requirements are
easily met by solutions of uranyl acetate, while uranyl nitrate solutions
are too acidic to be safe in this respect (Fig. 46). Several authors complain
that uranyl ions cannot be applied at a pH higher than 3.5, because upon
neutralization with alkali a precipitate of the corresponding diuranate will
appear. Investigations by de Boer et al. (1961) and Fuhs (1964) have shown,
however, that a complex of uranyl and ethylenediamine tetraacetate ions
in a ratio between one and two will furnish sufficiently high ionic activities
at pH values up to 8. Dangerously low pH values are found in solutions
of the salts of indium and lanthanum, and their use instead of uranyl salt
solutions is therefore discouraged.

Besides their conspicuous effect on nucleoid fine structure, uranyl ions
serve as an electron stain for nucleic acids. Association with the heavy
metal renders the DNA visible in the electron microscope. While several
authors have succeeded in demonstrating isolated DNA helices in the
electron microscope with the aid of uranyl acetate, the recognition of
single DNA helices in sections of moderate thickness is impossible due to
electron scattering in the embedding medium and generally reduced resolution. The fibrous material apparent in the sections consists of aggregates
composed of several helices each and related optical artifacts as depicted
in Fig. 42 (p. 97). Giesbrecht (1959 and later) contended that the thicker
fibres were preformed entities reflecting a tertiary structure of the DNA
("Molekularschrauben"). Fuhs (1965 b), however, in analyzing a pattern of
cross-sectioned fibres in a series of 200 Å thick sections was unable to trace
any of the fibrous aggregates from one section to another (Fig. 55). This
indicates that the pattern seen in the serial sections is due to random
aggregation of neighboring DNA fibres of the type shown in Fig. 42 *f*. In
the living cell, DNA helices most likely are completely separated from each
other in a pattern resembling Fig. 42 *a*.

4. Dehydration and Embedding

The effect of dehydration is instrumental in bringing about the changes
described in the preceding sections. The degree of shrinking during dehydration depends on the type of fixation and impregnation used. Results
obtained by Fuhs (1964) are summarized in Fig. 41 (p. 97). The type of

solvent used for dehydration depends on the requirements of the embedding medium and is of little importance for the preservation of nuclear detail. Dehydration occurs also during transfer of bacteria into water-miscible embedding media, the only difference being the replacement of the solvent by mixtures with water of the embedding medium.

A number of plastic embedding materials can be used with almost equal success for the preparation of ultra-thin sections of bacteria. There is no indication of any specific effect on nuclear fine structure. A slightly different appearance of nucleoids has been reported with a water-miscible resin (FUHS 1964).

5. Summary

From these considerations it appears that the RK-fixation provides an aspect of the DNA-plasm which is closest to natural, but fails to reveal the general orientation of the DNA fibres. Such orientation becomes more easily recognizable if moderate aggregation of neighboring DNA fibres is allowed to occur, as is accomplished by treatment producing the "fibrous" state (Figs. 37 and 38, p. 93). In view of this situation, illustrations on the following pages are mainly chosen from preparations showing the "fibrous state".

D. Nucleoid Morphology as Seen in Ultrathin Sections

Before entering detailed considerations on the fine structure and organization of nucleoids it appears advisable to correlate findings on nucleoid morphology with those obtained with the light microscope.

Some of the technical requirements for the correct preservation of nuclear fine structure need not be met for the correct preservation of nucleoid morphology. Artificial condensation of the DNA-plasm is permissible if it is realized that the nuclear body originally present corresponds to the entire area of the "nuclear vacuole" as outlined earlier (p. 90). Serial sectioning, however, is essential in obtaining reliable information on nucleoid morphology. A survey of a great number of single sections may reveal the general shape of the nuclear bodies, but still 80 per cent of the nuclear volume and more are excluded from analysis in any one cell observed.

Electron microscopic analysis confirms that nucleoids and nuclear bodies are elongated, prolate structures. This applies to the compact, centrally

Fig. 47. Action of dissolved salts on nuclear configuration in *Bacillus subtilis.* — *a* through *d,* light-micrographs of OsO₄-HCl-Giemsa preparations. — *a* control on salt-free medium. — *b* cells 5 minutes on salt medium (3% NaCl). — *c* 10 minutes on salt medium. — *d* 30 minutes on salt medium and showing reorganization of chromatin. — *e* through *g,* electron micrographs of ultrathin sections. Results are from investigations conducted prior to the introduction of the RK- and uranyl fixation techniques. Nucleoids therefore are in their artificially condensed state, and boundaries of nuclear areas *in vivo* are represented by the outlines of the nuclear "vacuoles". — *e* Cell grown in liquid tryptone—yeast extract medium deficient in salt (only 20 meq. of Na and 11 meq. of K were present). Cells were spun down and resuspended in a solution of osmium tetroxide in distilled water. Nucleoids are in their open or expanded state. — *f* Cell from same liquid medium with one percent of NaCl added. Osmium tetroxide (1 percent in Palade buffer), one part, was added to ten parts of the culture. Nuclear bodies show pronounced tendency to aggregate into central strands and masses. — *g* Cell grown in liquid salt-deficient medium as described under *e,* but resuspended after centrifugation in 3 per cent NaCl solution for 10 minutes. Then spun down again and resuspended in 1 per cent buffered osmium tetroxide solution. Nuclear bodies have contracted into single, centrally located masses. Upon resuspension in salt-free medium, the state of the nuclear material was reversed to that shown in *e.*
No magnifications given. — Photographs by Dr. R. G. E. MURRAY and Dr. C. F. ROBINOW. For more details the reader is referred to the paper by WHITFIELD and MURRAY (1956).

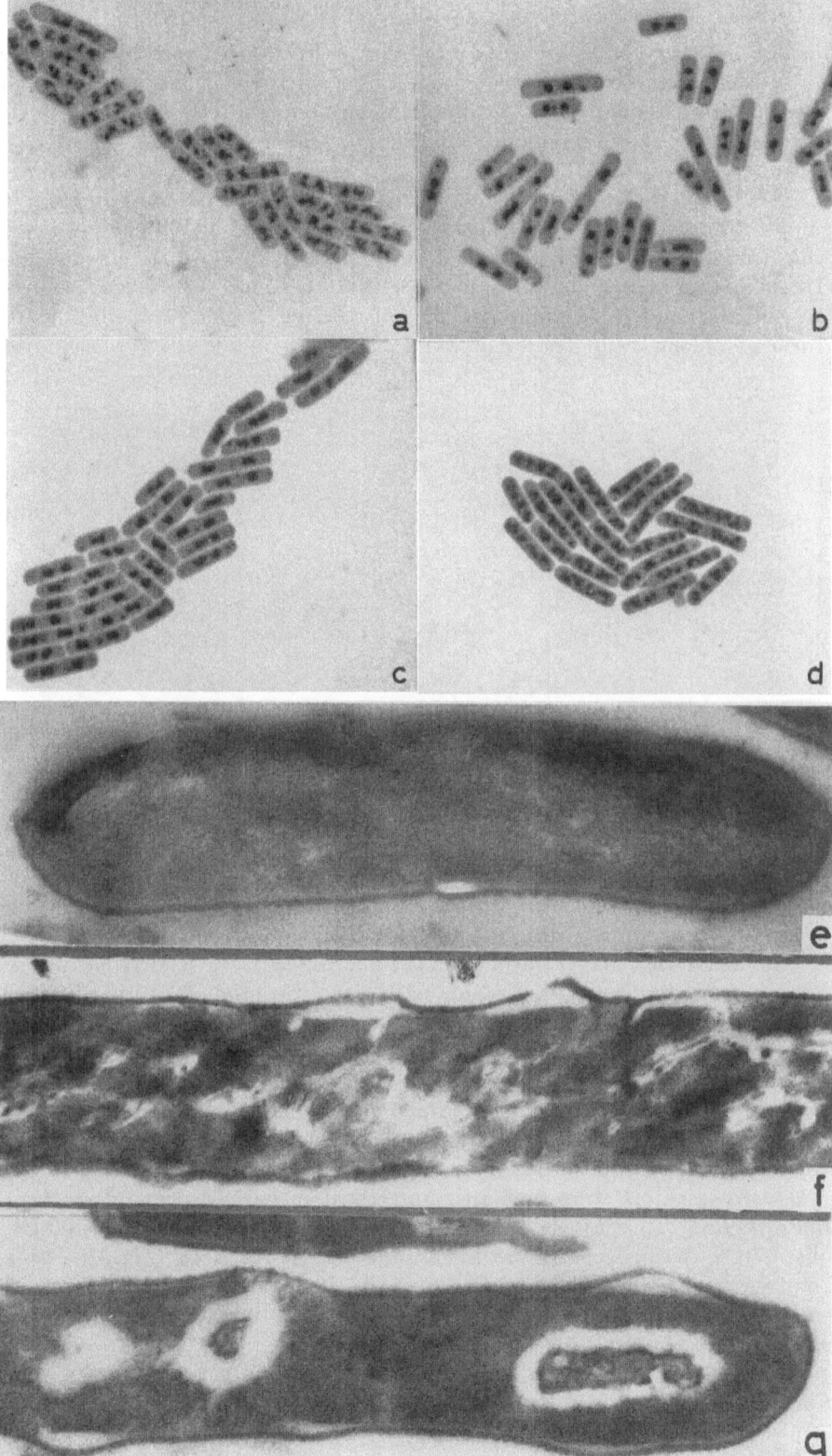

located nuclear bodies as well as to the ramifications found in the open or branched form. Evidence concerning the transition of the open into the compact form as dependent on the ionic environment (see p. 69) are partly based on electron microscopic observations (WHITFIELD and MURRAY 1956, PREUSSER 1959) (Fig. 47).

Several patterns of nucleoid morphology are conspicuously absent or under-represented in the electron-microscopic literature. This applies to the transverse-rod and V-shaped structures that have been described frequently in critical light-microscopic investigations. The electron-microscopic literature is abundant in demonstrations of nuclear bodies in the form of oblong axial bodies or axial filaments. While the representation

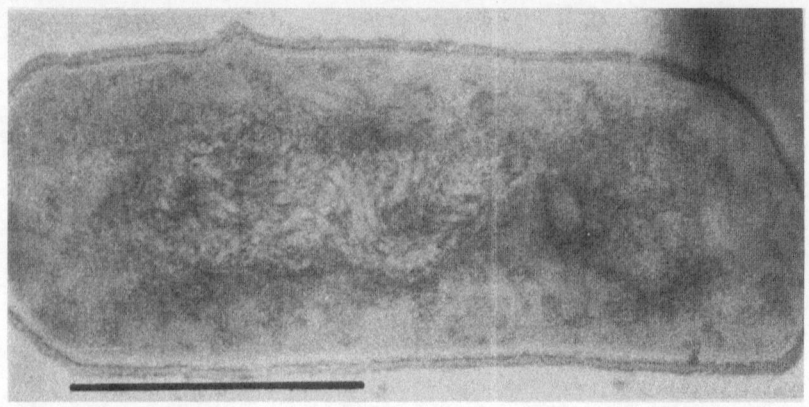

Fig. 48. *Bacillus subtilis*, from exponential growth phase in nutrient broth. Cytoplasm rich in RNA, particularly around nuclear area. Fibrous state of DNA plasm. — Scale marker 0.5 μ.

of transverse rods and V-shapes requires that the cell is sectioned under a particular angle, a requirement which is not easily met, the abundance of rounded forms may also be an artifact caused by extended fixation in osmium tetroxide (see p. 60). A survey of the electron-microscopic literature reveals a conspicuous lack of critical comparative work in this area. Another possible explanation is the under-representation in the electron-microscopic literature of bacteria from the lag and exponential phases. As pointed out earlier, compact or axial nuclear bodies tend to become predominant in the later phases of the growth cycle. The preference for cells from the declining or stationary phases of growth is understandable, since many investigators were primarily interested in basic aspects of nucleoid fine structure. The pattern of DNA fibres in rapidly growing cells is largely obscured by messenger and probably other species of RNA (Fig. 48), while "clearer" and undoubtedly more impressive illustrations are obtained with stationary cells. The preference for agar plate cultures adds another possibility of explaining the observed discrepancy.

The rounded nucleoid as a result of chloromycetin action has been mentioned earlier (Fig. 14, p. 48).

The open or branched form of the bacterial nuclear body has been shown in several cases to be due to ramification of a single coherent nuclear structure (Ryter 1960, Fuhs 1965 b) as can be expected if the structure represents a single nucleoid. The branches were found to be linear extensions or closed loops.

E. Nucleoid Fine Structure

Strictly spoken, the term "nucleoid fine structure" is misleading. It should be recalled from the introduction that the nucleoid is the genophor as observed in the light or electron microscope. As such, the nucleoid merely represents a defined portion of DNA-plasm. It should also be recalled from the preceding chapter that presently available techniques do not permit to trace a single DNA helix in ultrathin sections. Even if resolution were satisfactory, superposition of individual fibres would present a serious problem. This also implies that it will be difficult to recognize excursions of a single DNA fibre from the bulk of the DNA (the DNA-plasm) into adjacent cytoplasmic regions. We are left therefore with the problem of tracing the genophor from the general directional tendencies of the DNA fibres as seen in single or serial sections.

Experience shows that the use of serial sections which include all of the nuclear material (if possible, the entire cell) is even more imperative for the elucidation of nucleoid fine structure than it is for the investigation of nucleoid morphology. Attempts to substitute thick single sections for complete series (Giesbrecht 1959 and later) must be considered a failure, since the micrographs become overcrowded before they even cover one half thickness of an average nuclear body. Also experiments with stereomicrographs of thick sections and stereophotographs of superimposed micrographs of serial sections failed to indicate an easy road to the visualization of the organization of an entire nuclear body (Fuhs, unpublished results).

Any micrograph of a sectioned nuclear area that provides valid information according to the criteria which are discussed in the previous section reveals two basic principles of nucleoid organization:

1. The bulk of the DNA fibres exhibits a pronounced parallelism, and
2. The DNA fibres are quite evenly distributed throughout the DNA-plasm.

The number of DNA fibres that may be arranged parallel to form a bundle-like structure is neither constant nor limited. In some instances, the entire nucleoid appears to consist of a single bundle of parallel fibres the axis of which coincides with the long axis of the nuclear body (Fig. 53). In this case, the number of DNA helices penetrating any cross-section of a nucleoid will equal the length of the genophor divided by the length of the nuclear body representing a single nucleoid. In *E. coli* and *B. subtilis* this number is close to one thousand. In high-resolution electron micrographs of cross-sectioned bundles (e.g. Fig. 49 a) the approximate number of molecular fibres penetrating a given area of the section was determined and was found to agree with estimates of DNA content per cell as related to the nuclear volume (Fuhs 1965 b). From these determinations and from the degree of shrinking during dehydration (p. 97, Figs. 41 and 42) it appears that the average interhelix distance is approximately 150 Å.

Apart from this general principle, nucleoid structure is quite variable. The variety of patterns observed is greater than expected on the basis of changes in the plane of sectioning and changes during the cell cycle. On the other hand, the cell cycle has little if any conspicuous effect on nucleoid fine structure and on the geometrical pattern of the DNA helices. From these observations one may conclude that

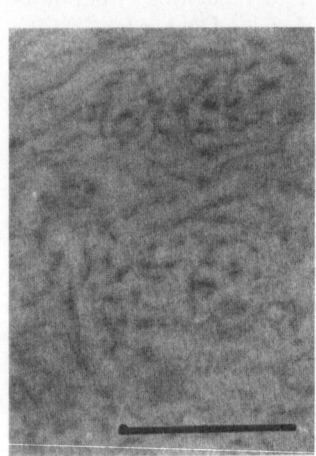 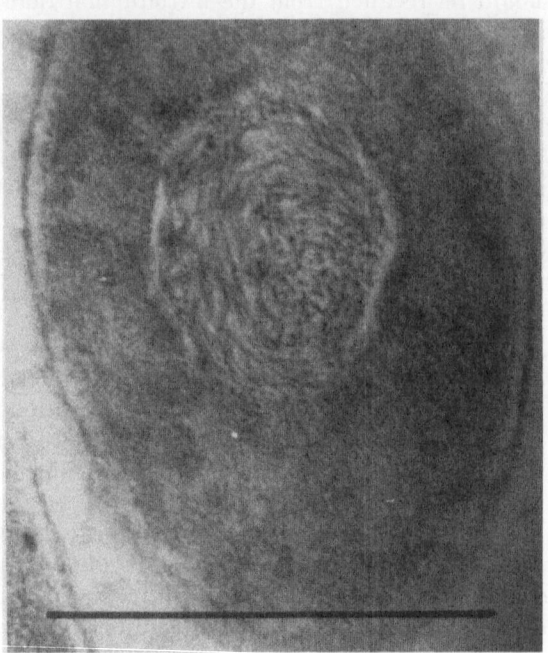

Fig. 49 a.   Fig. 49 b.

Fig. 49 a. Cross-sectioned bundle-like arrays of DNA fibres at higher magnification. Fibrous state of DNA plasm. Scale marker: 0.1 µ. — Reprinted from Fuhs (1965 d).
Fig. 49 b. Cross-sectioned cell of *Bacillus subtilis*. Two bundles of DNA fibres wrapped around each other. Fibrous state of DNA plasm. Scale marker 0.5 µ. Reprinted from Fuhs (1965 d).

(1) the cell cycle in bacteria is not accompanied by a temporary condensation of nuclear material as it occurs in the chromosomes of higher organisms, that

(2) nucleoid replication and division are processes requiring little geometrical rearrangement and probably are very simple separation processes which mainly occur on the molecular scale and therefore escape microscopic observation, and that

Fig. 50. *Bacillus subtilis*, cell from exponential phase of growth in aerated 1 % tryptone broth (no salt added) at 30° C. Open form of nucleoids, cytoplasm rich in RNA. Fibrous state of DNA plasm permits recognition of direction of DNA fibres which is in the direction of the various branches of the nucleoid. The cells apparently contain two nucleoids, each in a state of division. The individual nucleoids are V- or Y-shaped with diverging branches and their united parts close to the ¼ and ¾ marks of the long axis of the cell. There is a similarity with the cells shown in Fig. 35 and 52 which have one nucleoid in a more advanced state of division or with its division already completed. Scale marker represents 0.5 µ.

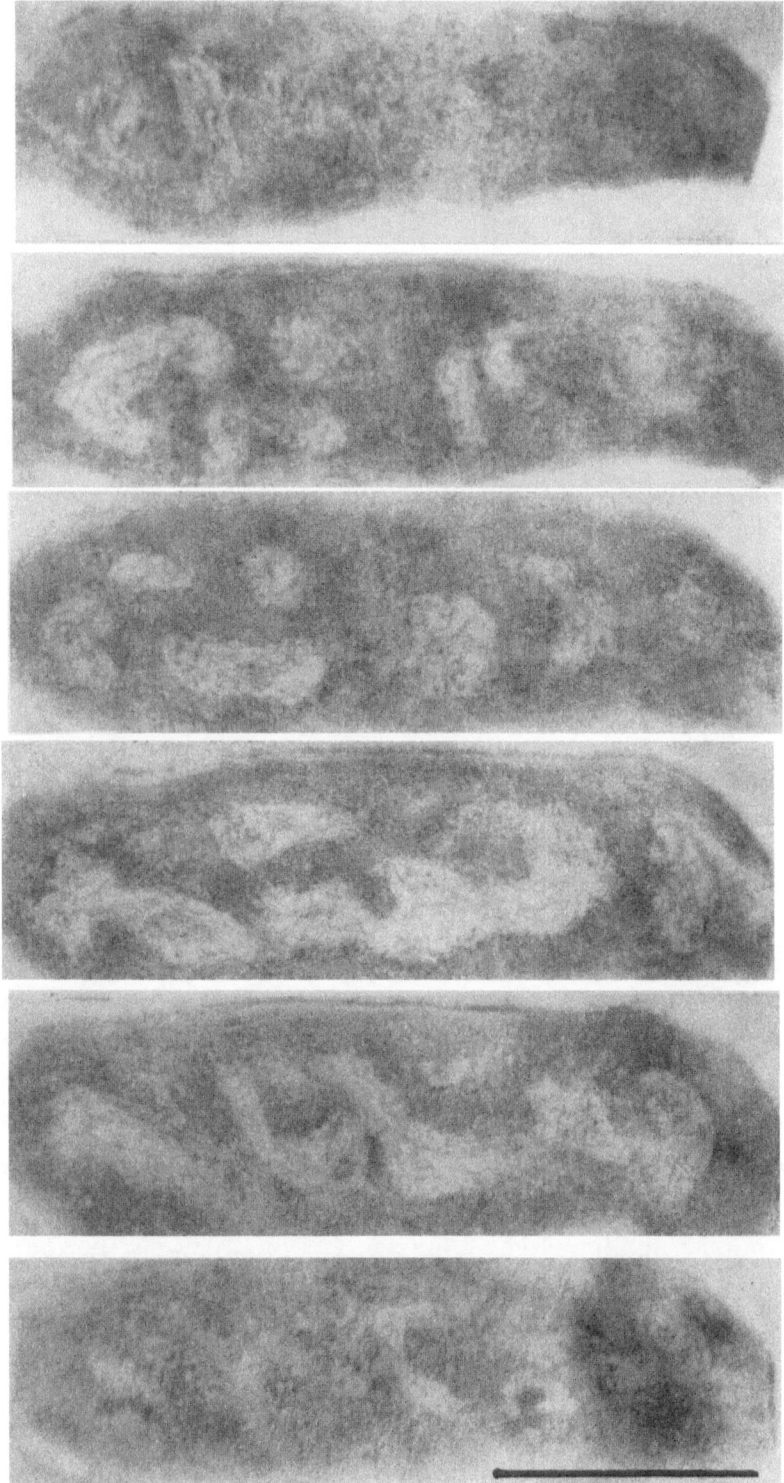

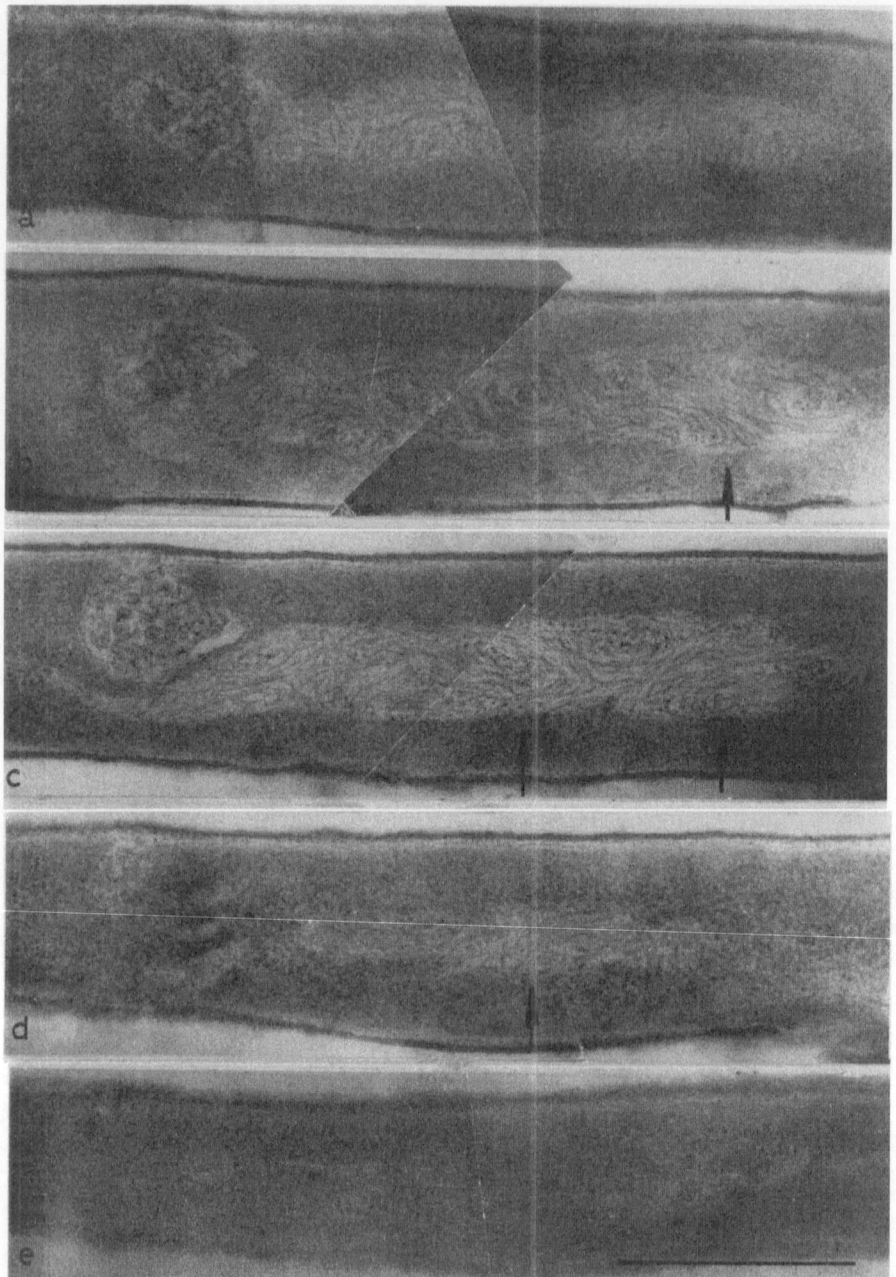

Fig. 51. *Bacillus subtilis*, fibrous state of DNA plasm, serial sections. DNA fibres in different planes of sectioning oriented in different directions (arrows). Scale marker is 0.5 μ. In part reprinted from Fuhs (1965 d).

(3) the nucleoid as a geometrical structure is flexible to a certain extent. permitting random variations in its general outlines which do not interfere with its inherent order which is topological rather than geometrical (Fuhs 1963 a).

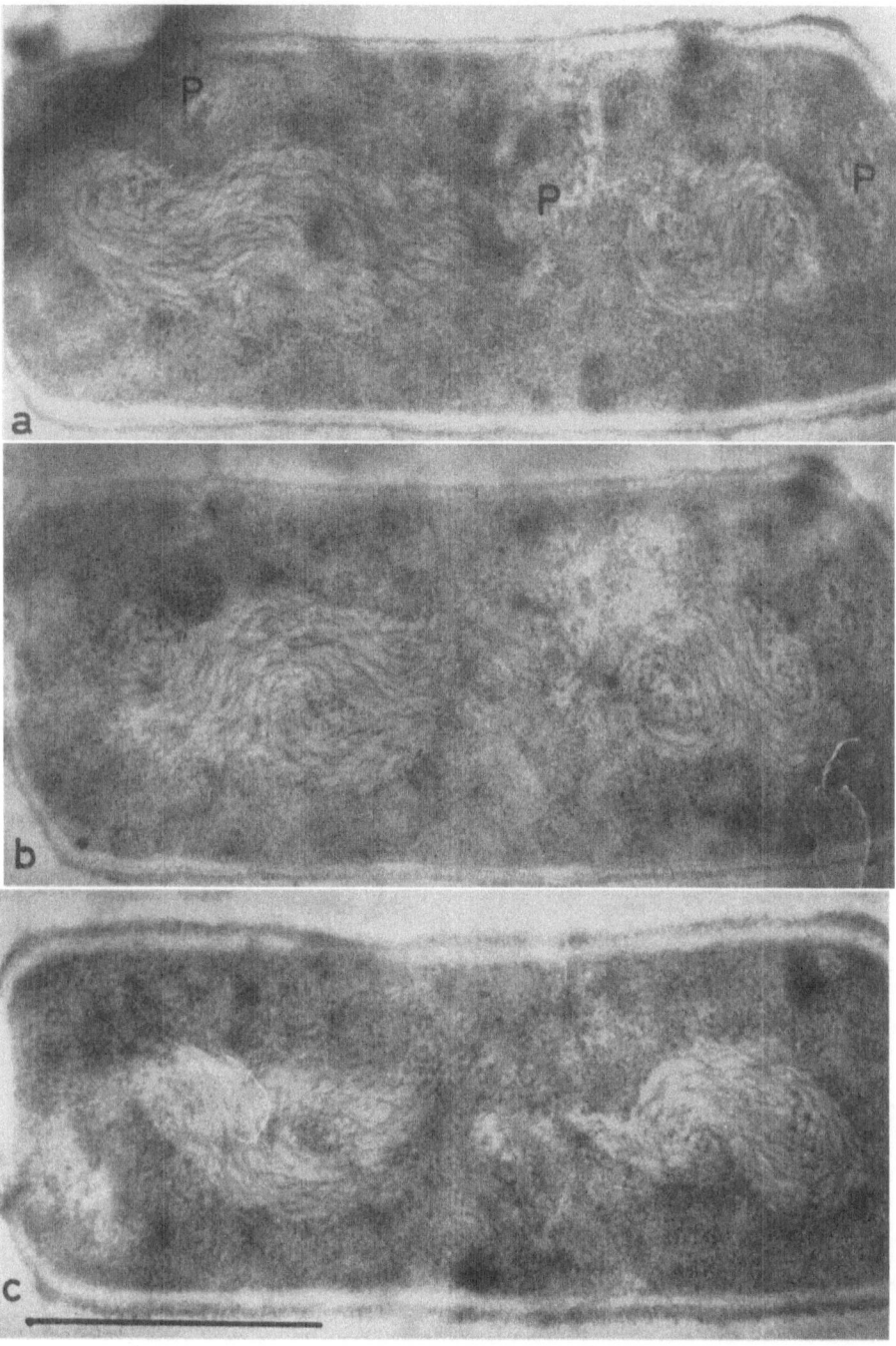

Fig. 52. *Bacillus subtilis*. fibrous state of DNA plasm, a short series of thicker sections. Cell shows two nucleoids which apparently have just formed by division. Each nucleoid apparently is attached to the central plasmalemmosome (*P*), proceeds towards the pole of the cell, turns backward, and ends near the center where it ends without an apparent attachment site. Reprinted after FUHS (1963a) with permission, Pergamon Press. Compare with Fig. 35. Scale marker: 0.5 μ.

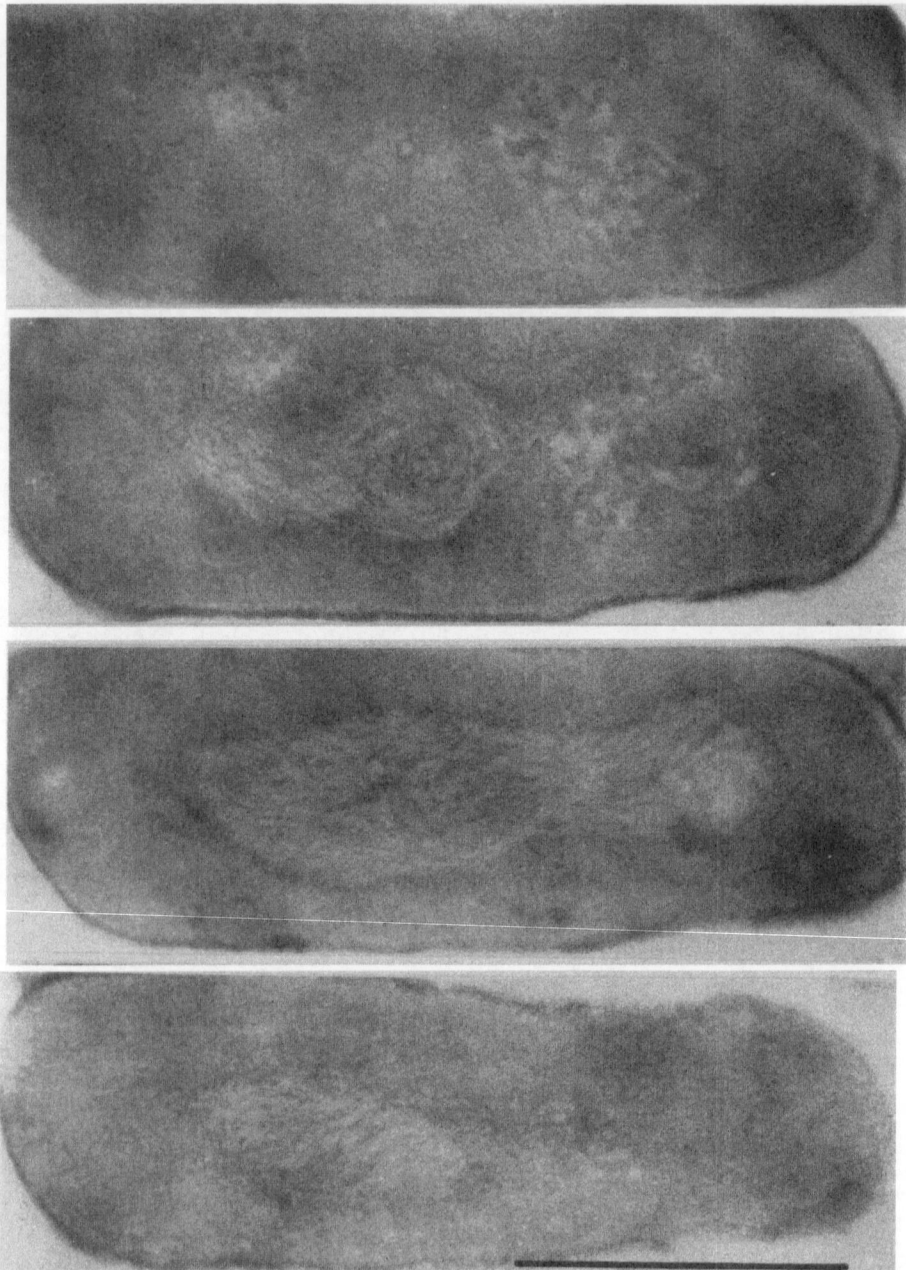

Fig. 53. *Bacillus subtilis*, fibrous state of DNA plasm, serial sectioning technique. Cell contains nucleoid in state of replication/division. Central whorl and branches extending to either side (and probably folded backwards) suggest pattern similar to that in Figs. 35, 50 and 52. Scale marker 0.5 µ.

In rod-shaped bacteria, the major part of the fibrous DNA material is oriented more or less parallel to the long axis of the cell, and the same holds for the majority of the branch-like extensions of a nuclear body in its "open" state. Most sections of such cells show several branches which in turn consist of numerous molecular fibres each (Fig. 50). In the compact state, the nuclear bodies, as a rule, consist of more than one such bundle; the individual bundles become apparent when they run under different angles with respect to the plane of sectioning (Fig. 49 b). Quite frequently, two bundles are wrapped around each other and form a cylindrical axial nuclear body. Quite complicated arrangements are frequent (Fig. 51), while the simpler patterns are quite rare (Figs. 52 and 53, see also Tomasz et al. 1964). The principal similarities between the open and compact states of nuclear organization explain the ease with which transition occurs as a result of changes in the ionic environment (p. 69).

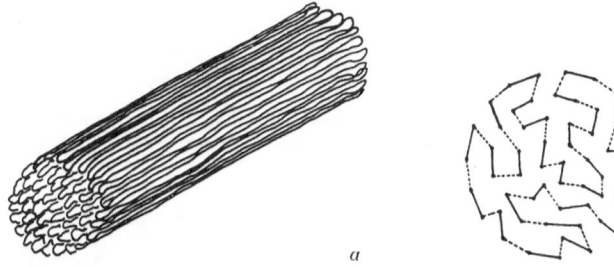

a   *b*

Fig. 54. Diagrammatic representation of bacterial nucleoid in its simplest form of geometrical organization, modified after Kellenberger (1960). Model is simplified in that actual number of DNA fibres is close to one thousand. — a Perspectivic view of model. — b Idealized cross-section of model showing connections between fibre segments in an arrangement that will permit tangle-free unreeling of the model and buildup by "stacking" of subsequent turns. Bonds at one pole are drawn as solid lines, bends at opposite end are dotted. — Fig. 54 a reprinted from Fuhs (1965 d).

In tracing the path of the "bundles" in serial sections it can be observed that a bundle turns backward at one of the poles of the nuclear body (Fig. 52, left pole). In such case the two bundles that are wrapped around each other over most of the distance obviously are part of the same structure. In Fig. 52, the nucleoid in the left half of the cell consists of a single bundle of fibres which originates in the center of the cell, proceeds towards the left pole, turns backward and in another plane returns to the central region of the cell. The nucleoid in the right half shows a similar arrangement. In the right nucleoid the possibility cannot be excluded that the entire structure represents a closed multistranded ring resembling a ∞, but in the left nucleoid a second turning point in the central part of the cell is definitely absent, and the bundle-like structure appears open-ended. From evidence cited early in this treatise it is apparent that the genophor is a closed structure. This rules out that the free end of a bundle consists of free-ending DNA molecular fibres. One must assume therefore that at such points adjacent DNA fibres are interconnected by small loops which escape detection. In this way the circular bacterial genophor would appear converted into a nuclear body by folding it back and forth into a bundle-like structure. This model was initially suggested by Kellenberger (1960) and modified by Fuhs (1965 d). The modifications consisted in the replacement by small loops of the initially proposed non-DNA linking groups, and in the "untwisting" of the model, since the twisted appearance was found to be due to the wrapping around each other of several bundle-like structures (Fig. 54).

Very simple nuclear patterns resembling Fig. 54 were observed in *Diplococcus pneumoniae* (Tomasz et al. 1964) and in protoplasts of *Bacillus*

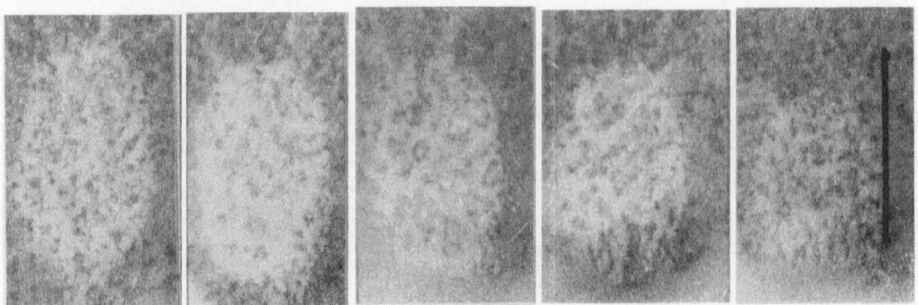

Fig. 55. *Bacillus subtilis* (T⁻ Romis), protoplasts, prepared by L. G. Caro (Dept. Biophysics, Univ. of Genève) with lysozyme in M 9 medium supplemented with Ca⁺⁺, Mg⁺⁺, sucrose and versene. Fixation in formalin followed by RK technique, embedded in methacrylate. — Series of transverse sections of nucleoid. — Scale marker is 0.25 μ.

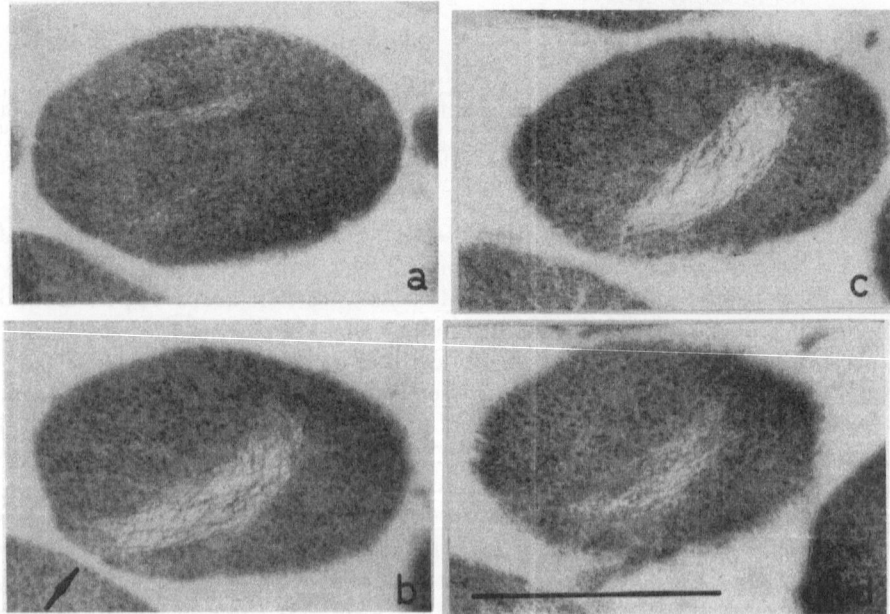

Fig. 56. *Bacillus subtilis* T⁻ protoplast (see legend to Fig. 55 for details of preparation). Serial sections. Nucleoid consisting of a single parallel array of DNA fibres and a narrow, tail-like extension (a). Arrow indicates possible site of attachment to membrane. Scale marker 0.5 μ. Reprinted from Fuhs (1965 d).

subtilis (Fuhs 1965 d). In these forms the nuclear bodies consist of single bundles of parallel fibres (Figs. 55 and 56). Contrary to other bacteria and normal *B. subtilis* cells, the nuclear bodies look very much alike, and division stages are readily discernible (see below). The existence of such nucleoids suggests that the more complicated types are derivatives of the simpler ones, and that the characters that distinguish the more complex

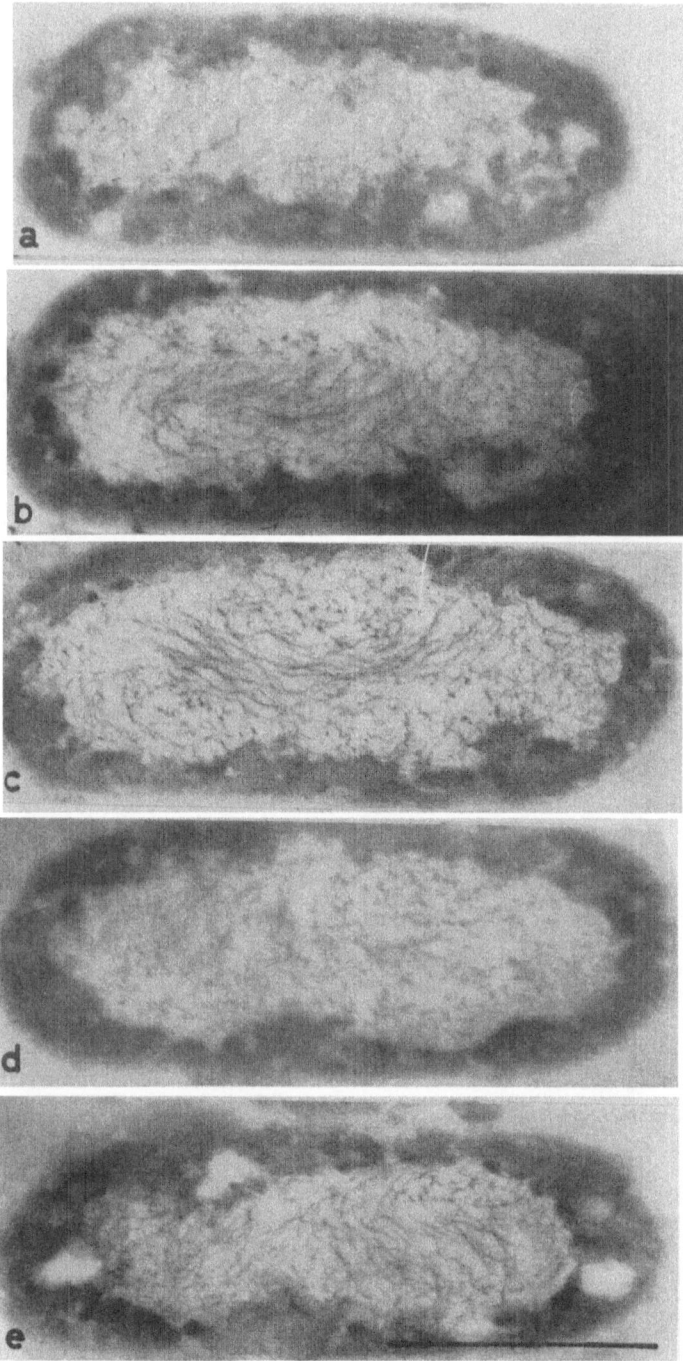

Fig. 57. *Escherichia coli* strain 15 T⁻A⁻U⁻, fibrous state of DNA plasm. Serial sections, section thickness 650 Å.
Scale marker 0.5 μ.

types from the simpler ones probably are non-essential for the proper functioning of the nucleoid and to a certain extent may be the result of random displacement. While more investigations are needed to clarify this point in greater detail, it appears safe to state that a common principle is materialized in both aspects of nuclear organization, and neither the complexity of the one nor the simplicity of the other interferes with the processes of nucleoid replication and genome segregation.

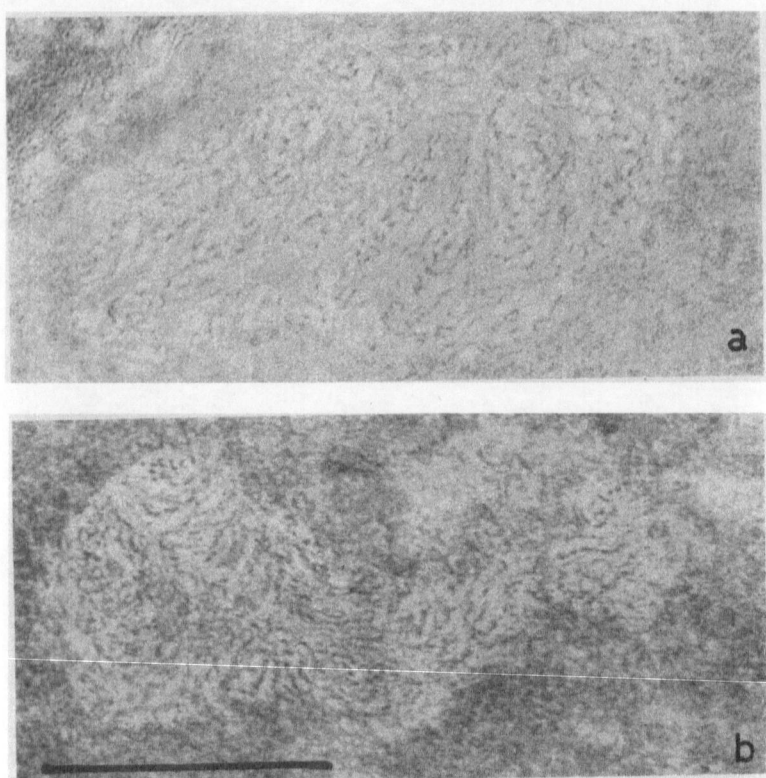

Fig. 58. Ultrathin sections of germinating spores of *Bacillus subtilis*. Fibrous state of DNA plasm. — *a* Longitudinal section through nuclear body. The pattern resembles that seen in the chromosomes of dinoflagellate if sectioned parallel to their longitudinal axis. — *b* Nuclear body as seen in oblique section through a germinating spore (direction of sectioning was judged from the appearance of the cell wall which is not shown in the Figures). In the left half of the nucleoid the featherlike structures become apparent which are also seen in obliquely sectioned chromosomes of dinoflagellate. — Scale marker represents 0.25 μ. Micrographs by C. R. and J. Marak, obtained through the courtesy of Dr. C. F. Robinow.

Contrary to earlier expectations there appears to be no basic difference in fine structure and organization of nucleoids from Gram-negative and Gram-positive bacteria (Fig. 57, see also Ryter and Piéchaud 1963, and Tikhonenko et al. 1964). Some of the difficulties encountered by other workers may be due to failure to establish proper conditions for the RK-reaction in a variety of bacteria in different phases of the growth cycle.

The pattern shown in Fig. 58 is occasionally observed in oblique sections of sporeforming bacteria. It occurs more frequently in germinating spores and is referred to as "featherlike structure" (Robinow 1962, Giesbrecht

1962). This pattern has aroused considerable interest, since it resembles an arrangement found with great regularity and greater uniformity in the chromosomes of dinoflagellates (GRELL and WOHLFARTH-BOTTERMANN 1957, KELLENBERGER 1963, DE HALLER et al. 1964, GRELL and SCHWALBACH 1965). On the basis of micrographs published by DE HALLER et al. the pattern in all likelihood is derived from the one shown in Fig. 54 (p. 113) by a contraction or compression of the bundle in its longitudinal axis. As a result the individual fibres would follow a helical path the pitch of which equals the repeating unit seen in the micrographs. If the fibres stay in phase, the resulting pattern may resemble that shown in Fig. 58. This explanation as any other thus far offered remains subject to verification by serial sectioning. This statement in particular applies to model concepts advanced over a number of years by GIESBRECHT. The reader is referred to a 1965 paper for his most recent revision of the model. Earlier concepts proposed by this author suffer from the disadvantage of being incompatible with the microscopically observed splitting pattern of dinoflagellate chromosomes (DODGE 1963, 1964 a) and with regard to their validity in bacteria were rejected following critical examination (FUHS 1965 b). From Fig. 59 it is apparent that in bacteria the pattern lacks the rigidity regularly observed in dinoflagellate. Unfortunately little is known of the molecular organization of the dinoflagellate chromosome so that comparative examinations are of limited value.

F. Relationships between the Nucleoid and the Cytoplasmic Membrane

Some form of association between the nucleoid and the cytoplasmic membrane has been assumed to exist by several authors for several and at times different reasons. Cytological evidence favors such assumption, since points of contact between these organelles are observed quite regularly, especially if careful surveys are made using the serial sectioning technique. The frequent association of the nucleoid with the plasmalemmosomes has received particular attention. (The latter structures are invaginations of the cytoplasmic membrane [VAN ITERSON 1960, 1962, GLAUERT et al. 1961, GLAUERT 1962, RYTER and JACOB 1963, 1964, SEDAR and BURDE 1965, FUHS 1965 b]. The frequently used term "mesosome" is ambiguous and less descriptive, see EDWARDS 1962.)

Contact between nucleoids and plasmalemmosomes is strong enough to withstand physical strain if the membranous organelle is extruded during protoplast formation. The attached area of the nucleoid is pulled towards the cell surface (Fig. 60, RYTER and JACOB 1963, 1964, 1966 a, RYTER and LANDMAN 1964, FITZ-JAMES 1964 c). FITZ-JAMES (1965) summarized evidence concerning the regularity of occurrence and possible significance of the membrane-nucleoid association.

RYTER and JACOB (1963) presented serial sections of B. subtilis cells which contained numerous plasmalemmosomes and frequent points of contact between these and the nucleoids. The micrographs seemed to suggest that plasmalemmosomes may serve as mechanical anchor sites for the nucleoid which are important for the separation of daughter nucleoids at division.

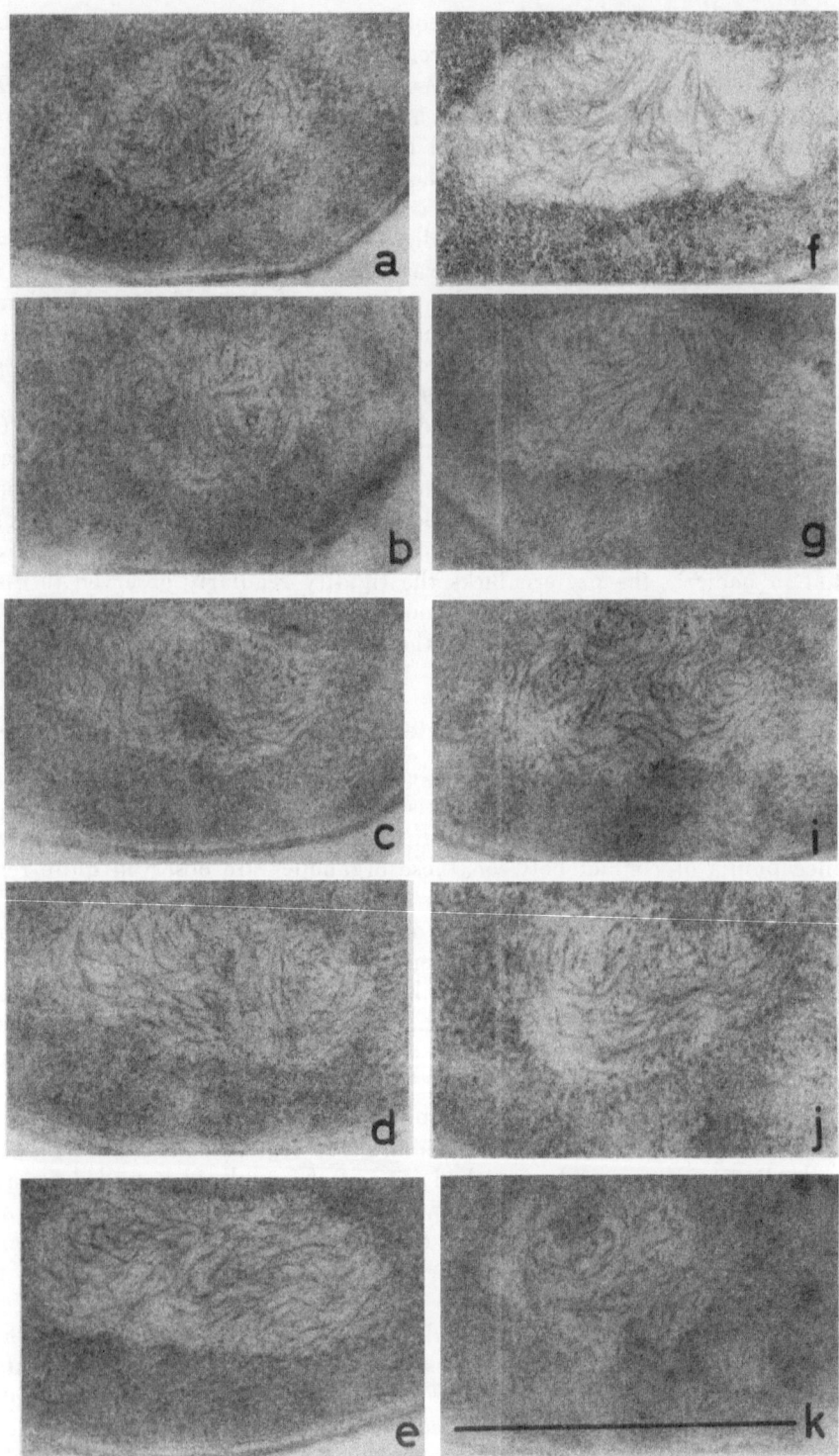

Fig. 59 a—k.

Fuhs (1965 b) in serial sections of a different strain of the same species showing a much smaller number of plasmalemmosomes per cell was unable to establish a similar correlation. This is explained by the apparent attachment of the nucleoid also to less conspicuous sites at the membrane (Ryter 1967) (Fig. 61),

In Gram-negative bacteria, points of contact between the nuclear bodies and the cytoplasmic membrane are always established without the interference of plasmalemmosomes, since these structures are rarely found in these bacteria (Ryter and Jacob 1966 a). This indicates that with respect to nucleoid attachment peripheral membrane sites and sites at the surface of a plasmalemmosome are equivalent. While the frequent association with

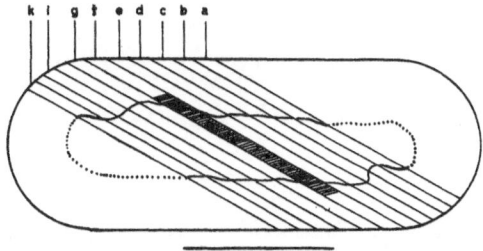

Fig. 59 *l*.

Fig. 59. Serial sections through nuclear body of *Bacillus subtilis*, fibrous state of DNA plasm. Section *d* shows the featherlike arrangement of DNA fibres. The other sections show that this pattern is not as rigid and regular as in the chromosomes of dinoflagellate where it extends through all planes of sectioning. Nor can this pattern be interpreted in terms of a single, supercoiled structure making up the entire nucleoid. Note that section *h* is not shown, since due to accidental superposition of two sections a clear micrograph could not be taken from this section. — *l*. The diagram shows a reconstruction of the sectioned cell derived from the outlines of the cell and the nuclear body in all sections obtained. Note that the series comprises all planes of the nucleoid over the range of section *d* (hatched) which contains the featherlike structures. — Both scale markers represent 0.5 μ.

a plasmalemmosome may be less than accidental, functioning of the site with respect to the nucleoid does not necessarily require its location on a plasmalemmosome. If we assume that the attached segment of DNA is an operon actively involved in the direct synthesis of membrane or cell wall material (p. 30), its function then may result in a temporary production of excess membrane or wall precursor material which in turn results in the appearance of a plasmalemmosome. In any case the accumulated evidence suggests that nucleoid segregation is under a more uniform and stringent control than the formation of plasmalemmosomes so that most probably the latter process does not control the former.

In order to discuss more fully the possible significance of membrane attachment we have to distinguish between two possibilities; first, the attachment to the membrane of a single segment or operon of the genophor, and second, the attachment of several or numerous segments which are separated by non-attached segments. We refer to these possibilities as single-point or multipoint attachment.

Single-point attachment is a phenomenon referred to earlier in this review in conjunction with mechanisms of genome replication and segregation (p. 32). Cytologically the passage of a single DNA fibre to and from the membrane through a layer of cytoplasmic material is difficult to trace, and micrographs by Ryter and Jacob (1966 a) and Ryter (1967) (Fig. 61) and the indication of a narrow channel in Fig. 56 are the currently available optimum, and even here the exact number of DNA fibres remains undetermined. The segment of the genophor associated with the membrane could be the origin of replication or replicator, or any other genetically defined site, or the "complex of replication", i.e. the replication fork. The former types would establish a physical connection between a defined segment of

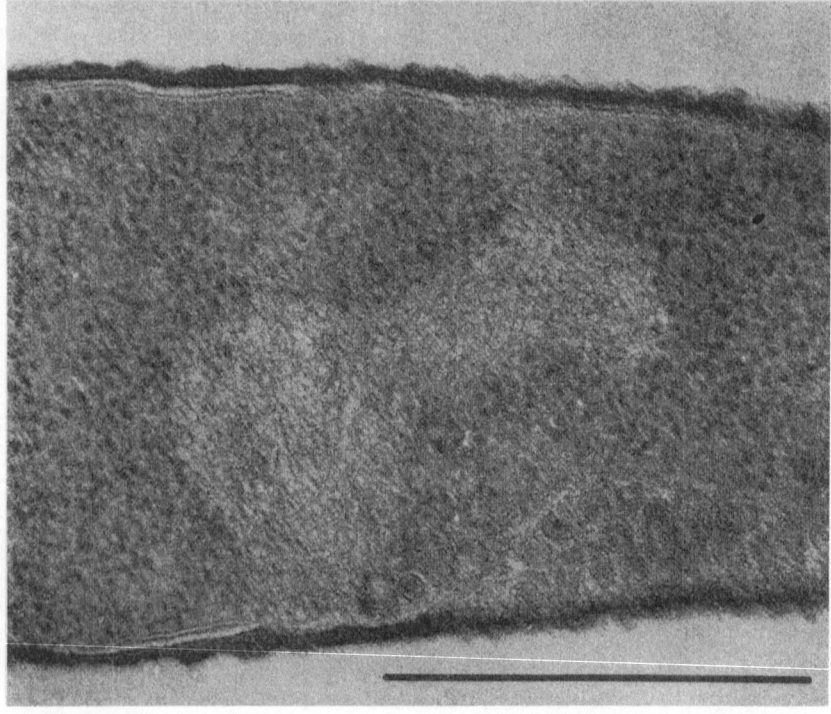

Fig. 60. Nucleoid-membrane association in *Bacillus subtilis* during protoplast formation. The plasmalemmosome shown at the lower right is in the process of being extruded, but contact with the nuclear body is maintained in the process, and the nuclear body is dragged towards the peripheral membrane of which the extruded plasmalemmosome forms a part. — Scale marker represents 0.5 μ. Reprinted from Ryter and Landman (1964).

the genophore and a segment of the membrane, while the latter type would cause the genophor to pass along that membrane site. Single-point attachment hardly will provide a mechanical holdfast for the entire nucleoid.

Multipoint attachment is the condition apparently referred to in some papers demonstrating a quite conspicuous zone of contact between a nuclear body and the membrane. Solid multipoint attachment involving a number of defined segments of the genophor could serve as a mechanical anchor for movements of the entire nucleoid but is likely to interfere with any of the proposed mechanisms of nucleoid replication.

Obviously some of the functions tentatively assigned to the nucleoid-membrane attachment site are mutually exclusive. A summary of the currently discussed possibilities is given in the following section.

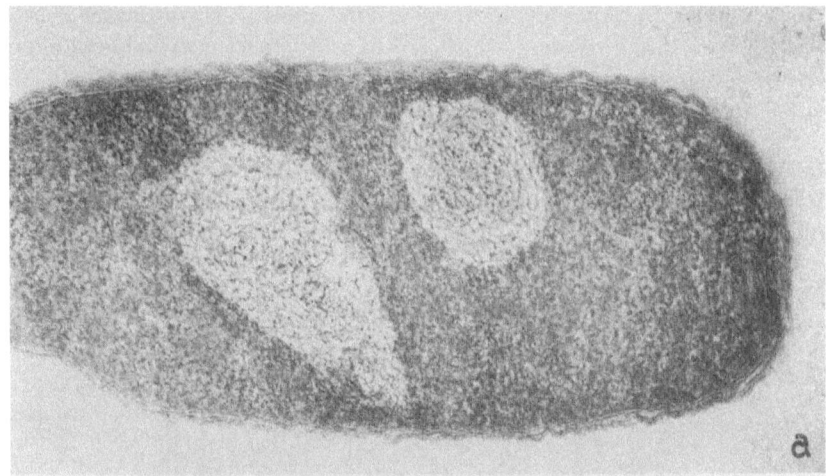

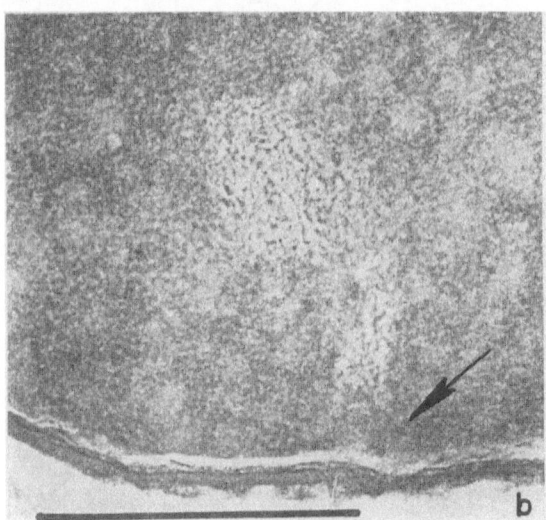

Fig. 61. Modes of nucleoid attachment to membrane. Attachment not mediated by plasmalemmosomes. — *a* Small area of contact in temperature-sensitive *Escherichia coli* 257. Magnification not given. Photograph by Dr. ANTOI-NETTE RYTER. — *b* Attachment mediated by a delicate intermediate structure (arrow), in cell of *Bacillus subtilis* in germination. Scale marker represents 0.5 μ. Reprinted from RYTER (1967).

G. Fine Structure of the Dividing Nucleoid

Spectacular changes such as the condensation of chromosomes which are observed in eucaryotic cells are never found in bacteria of Cyanophyceae. In the protocaryon, transcription of DNA, replication and genome segregation can proceed simultaneously and without apparent major reorganization. This suggests that DNA replication is accompanied by the segregation of the daughter helices into two easily separable masses of nuclear material. Additional requirements for any model of nucleoid replication and division are also that unfolding can produce appearances as shown in Figs. 4 and 5, (p. 13 and 14) and possibly that a new round of replication can be initiated before the preceding one has been completed (multifork replication, p. 126). Especially the latter process which is known to produce viable progeny is difficult to visualize without assuming that daughter helices of DNA are separated without delay. In this picture, nuclear bodies such as shown in Fig. 51 (p. 110) are assumed to consist of several independent nucleoids which may separate immediately prior to or even during cell division (Fig. 62) (CHAPMAN 1959, 1960, CONTI and GETTNER 1962, GRULA and SMITH 1965).

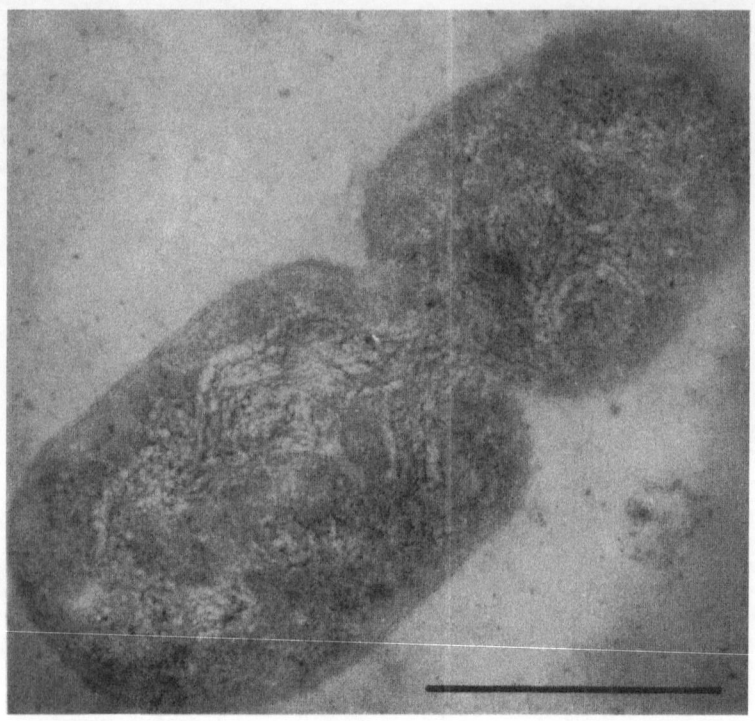

Fig. 62. *Escherichia coli*, cell division in progress before nucleoid division is completed. Fibrous state of DNA plasm. Scale marker: 0.5 µ.

As mentioned earlier, molecular and cytological evidence appears to converge into a consolidated picture of nucleoid structure which except for some minor details is as proposed by KELLENBERGER (1960, Fig. 54, p. 113). This author also attempted to design a model of nucleoid replication which was in agreement with the established fact of a circular genetical structure. It was assumed that the basic organization of the nucleoid could be described as that of an endless thread of DNA folded forth and back on the surface of an imaginary cylinder (Fig. 63 a). Such model could undergo replication without disruption of the DNA thread by producing a second, concentric cylinder. Originally it was thought that the daughter structure might form at the outside and would separate by being "peeled off" the mother cylinder. This proved to be an unlikely event, since still one double break was required for the separation of the daughter structures, and because sideways separation of real nucleoids which are solid rather than hollow cylinders would involve a considerable chance of tangling.

A workable model can be derived from the original one, if some minor modifications are introduced. As a first step let us assume that the concentric cylinders, once completed, would slip apart in a longitudinal direction (Fig. 63 b). This model could operate also in the case of a collapsed, massive cylinder, the cross-section of which would be as shown in Fig. 54 b, but still some additional geometrical requirements had to be met to avoid tangling during segregation. The model also implies that replication and segregation are separate phases of the cell cycle. Both predictions are not borne out by observations. Examination of patterns as shown in Figs. 56, 64, and 65 suggests that replication and segregation may indeed occur simultaneously in that each segment of one of the daughter structures is extruded from the bulk of the nucleoid immediately after being formed, and in a longitudinal direction. Fig. 56 shows a *B. subtilis* protoplast containing a single nucleoid. In the uppermost section, a narrow tail-like extension of the nucleoid can be seen. In Fig. 64, the nucleoid is bipartite or roughly V-shaped, the branch shown in the upper section being of less than normal diameter. Fig. 65 shows a protoplast with two nucleoids of approximately equal size which are arranged in a V-like pattern resembling that in the preceding illustrations. There is little doubt that the topography of the cell was significantly changed during protoplast formation and also that protoplasting contributed to the uniform appearance of the nuclear patterns. On the

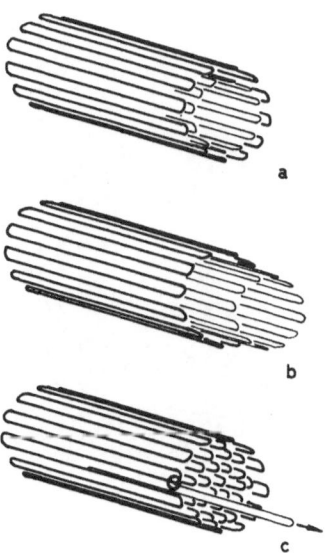

Fig. 63. Derivation of models of nucleoid replication.—*a* Topological model showing genophor arranged on the surface of an imaginary cylinder.—*b* Topological model showing separation of daughter cylinders. — *c* Geometrical model showing nucleoid in the process of replication and immediate segregation of replicated part. Tangling is avoided if arrangement of fibres in the nucleoid is as shown in Fig. 54 b, p. 113.

other hand, the arrangement in the protoplasts is not significantly different from that found in some intact cells (Tomasz et al. 1964, also Fig. 53, p. 112), and the argument that excessive breakage of the genophor does not occur during preparation applies also to the protoplast specimens shown here. It therefore appears that the appearance is altered in the sense of relaxation rather than disruption of nucleoid fine structure. Contrary to similar preparations, RK-conditions were not fully met during fixation which in this case contributes towards the clarity of the fibrillar pattern.

The illustrations suggest that a daughter nucleoid is built up gradually from one pole of the existing nucleoid. The material incorporated into the new nucleoid in all probability consists in one of the DNA daughter structures while the other daughter structure gradually replaces the mother structure.

While it is definitely not possible to develop at the present time a definite model of nucleoid structure and replication, both molecular and cytological evidence presented in this review does suggest that room for speculation has become very limited. To illustrate this situation we shall investigate

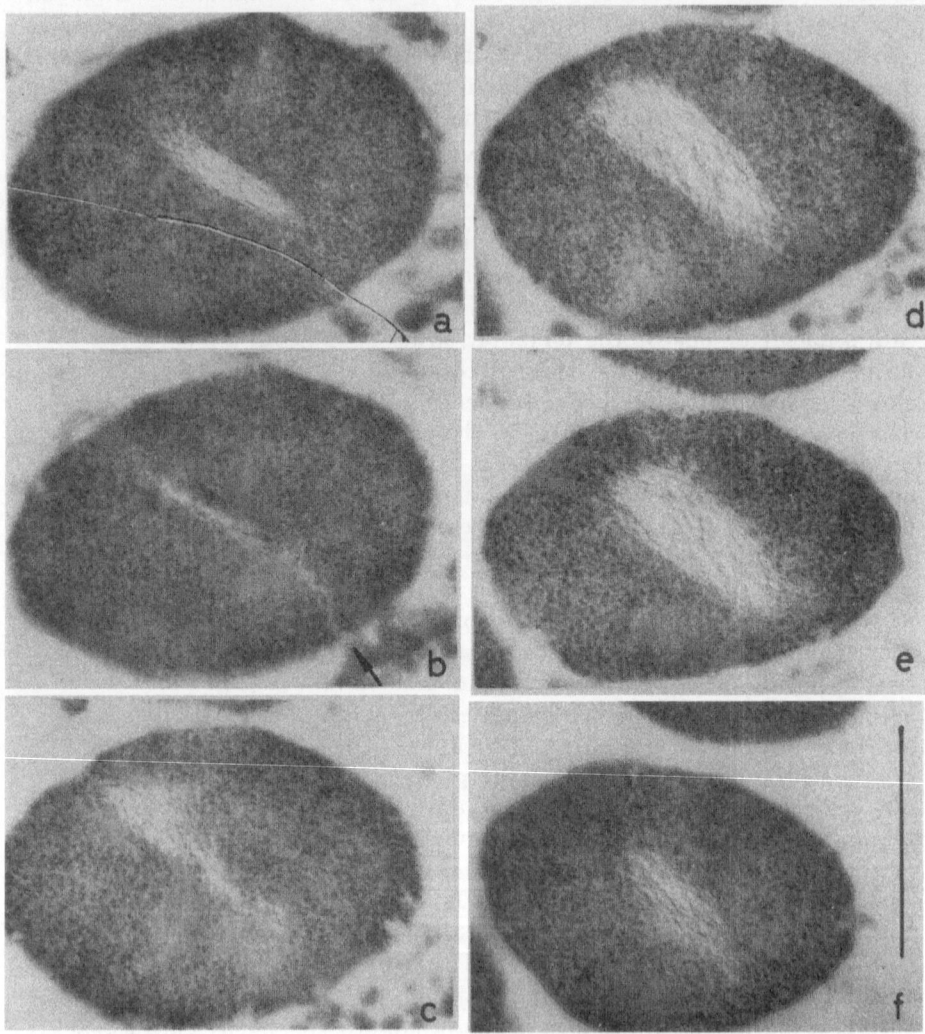

Fig. 64. Protoplast of *Bacillus subtilis* T⁻ (see legend to Fig. 55 for details of preparation). Serial sectioning technique. Nucleoid consists of two bundle-like arrays of DNA fibres which are situated in the planes of Figures *a* and *b*, and *c* through *f* respectively. The two bundles are unequal in size and arranged in the form of a "V". Arrow indicates possible site of attachment to membrane. Scale marker: 0.5 μ.

further the possibilities of the modified KELLENBERGER model shown in Fig. 63. The suggested sequential mode of replication coupled with segregation is illustrated in Fig. 63c.

Any attempt to define this model more clearly will involve a commitment as to its moving or stationary parts. This is also with reference to possible membrane attachment which may involve either the origin of replication, the replication fork, an operon, or possibly several of these— plus the possibility of an involvement of the episomal species. Two major

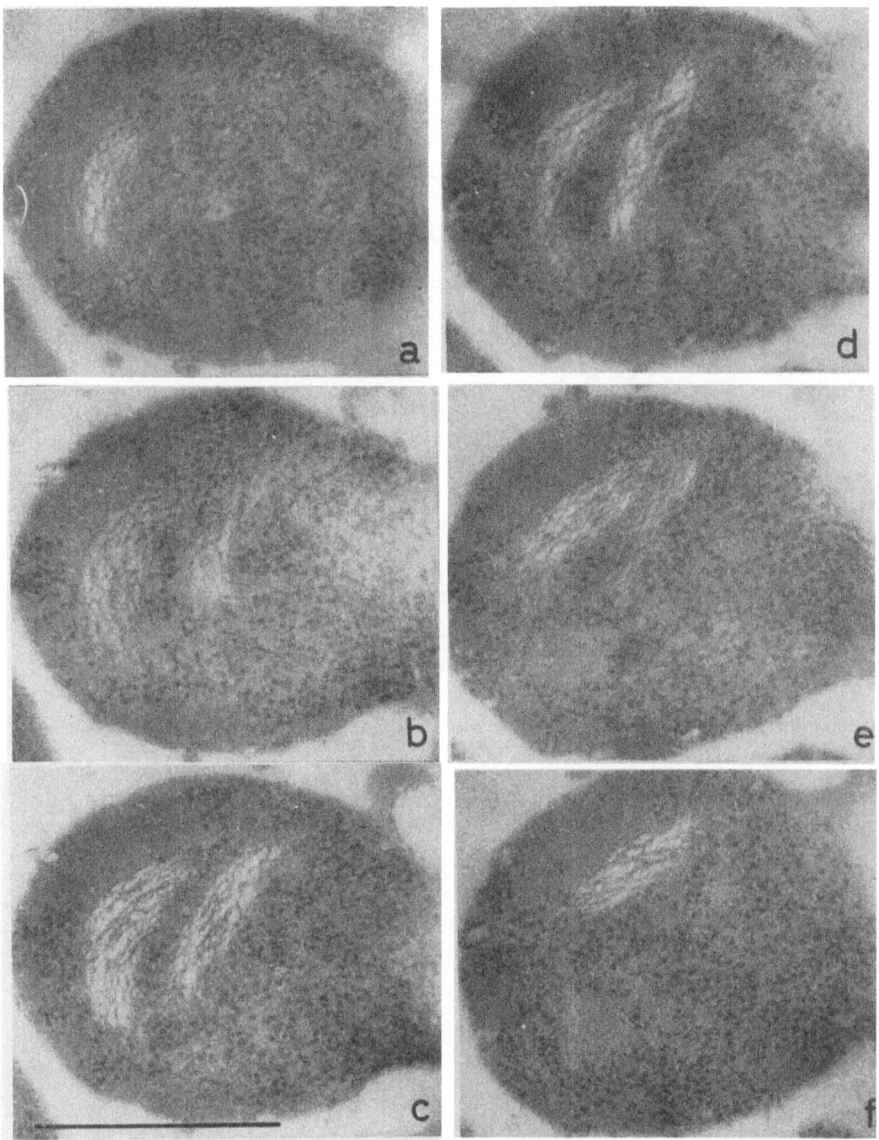

Fig. 65. Protoplast of *Bacillus subtilis* T⁻ (see legend to Fig. 55 for details of preparation). Serial sectioning technique. Nucleoid has completed the process of replication/division. Two nucleoids of equal size arranged in form of a "V". Scale marker represents 0.5 μ.

alternatives are selected and shown in Figs. 66 and 67. In Fig. 66, the nucleoid rests in place while being replicated; the replication fork moves along the stationary genophor. This would allow for attachment of the origin (as shown) or—with a slight modification only—attachment of any

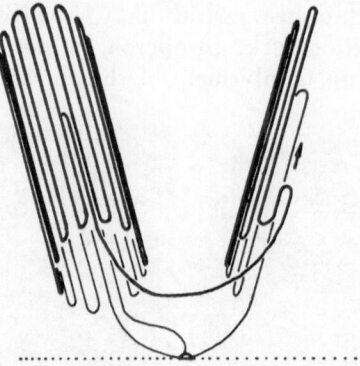

Fig. 66 a. Model of nucleoid replication/division based on the assumption that the origin is located at some fixed site.

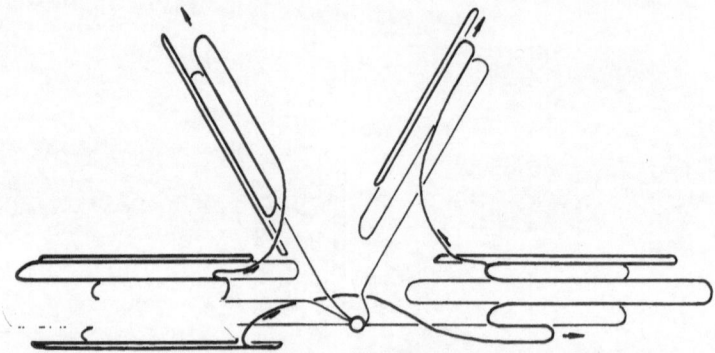

Fig. 66 b. Modification of the model shown in Fig. 66 a demonstrating the feasibility of premature initiation of a second replication cycle (multifork replication) in both daughter structures.

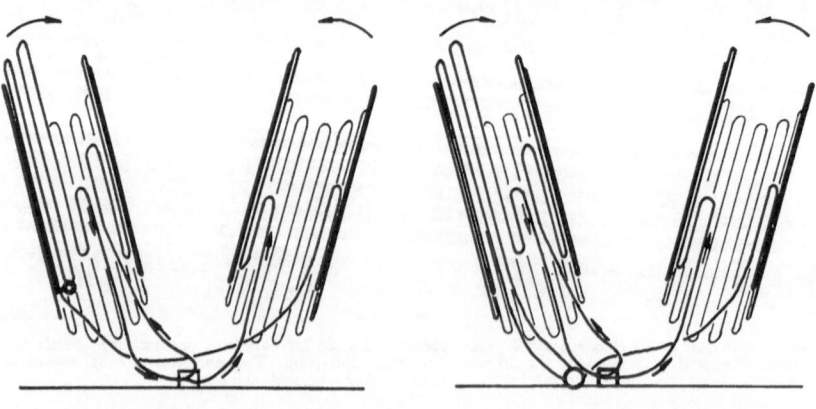

Fig. 67.   Fig. 68.

Fig. 67. Model of nucleoid replication/division based on the assumption by Jacob et al. (1963) that the replication fork is located at some fixed site.

Fig. 68. Model of nucleoid replication/division demonstrating the feasibility of both origin and replication fork being attached to some membrane site. It is understood, however, that such attachment is a temporary one.

other genetically determined site. Multifork replication is compatible with the model as shown in Fig. 66 b. The orientation of the new nucleoid with reference to the existing one is undetermined; it may be synthesized alongside the other, opposite to it, or under any angle producing a V-shaped arrangement, or the existing nucleoid may not be replaced by one of the daughter structures. Nor is the compact structure of any of the nucleoids an absolute requirement; they may exist in their "open" or branched conformation as well. Thus it appears that a rigid geometrical arrangement is not an indispensable prerequisite for the orderly replication and segregation of the bacterial nucleoid which is suggested by the multitude of patterns and configurations observed.

The representation of Fig. 67 is different from the one in Fig. 66 in that the genophor travels along the replication mechanism which then may be assumed to be in some fixed position. In discussing this model one may consider provision of a "driving force" versus the need of procuring a substantial amount of precursor material in a single location. Multifork replication is less easily accommodated. This model is in closer correspondence with that proposed by Jacob et al. (1963) (p. 30), but the proposed alternate mode of control by bacterial and episomal systems and the possible membrane attachment of the latter are still unsolved problems.

Finally it appears that both models shown may be combined in a third (Fig. 68) which contains the origin as well as the complex of replication in immediately adjacent locations which may be fixed. The bulk of the nucleoid would remain stationary as shown in Fig. 66, but the genophor would pass through the complex of replication in sequence and produce two daughter nucleoids. In any case, the attached site would be the last to be replicated. If segregation of the attachment site occurs earlier (see Ryter 1967), the attachment site could be neither the origin, nor the replication fork (see also Ryter 1968).

While it is impossible to arrive at more definite conclusions at the present time, the models as well as the electron micrographs presented earlier caution against postulating longitudinal fission in the traditional sense as the only feasible mechanism of division of prolate genetical structures (see p. 147 for a more detailed consideration of this point).

In order to elucidate nucleoid structure more fully, the continued improvement of histochemical and autoradiographic techniques (Malmon 1967) will prove extremely helpful.

H. The Association of Nucleoids with Cell Components Other than the Cytoplasmic Membrane

To a certain extent nucleoid morphology and fine structure are determined by other cell inclusions. Open and compact nucleoids may include lipid granules, polyphosphate granules, or isolated portions of cytoplasm. The presence of large quantities of storage materials will force the nucleoids to accommodate themselves in the remaining areas of the cell. This contributes towards the variability of the nuclear patterns.

The close association of nucleoids and lipid inclusions and the possible deformation of the former by the latter has been observed frequently (Stille 1937, Piekarski 1940, Lewis 1941, Burdon et al. 1942, Flewett 1948, Smith 1950, Delaporte 1950, Robinow and Hannay 1953). Giesbrecht (1957) mentions lipid-like but "unstainable" material as a nuclear core in HCl-Giemsa preparations of *B. megaterium* and studied the role of the lipid

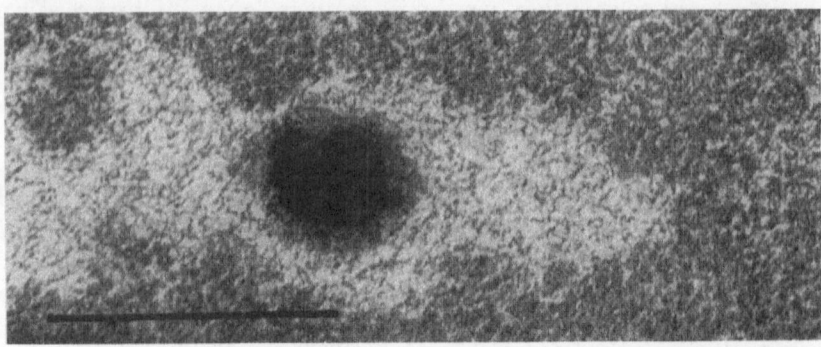

Fig. 69. Ultrathin section of *Myxococcus xanthus*. Incomplete polyphosphate granule surrounded by a narrow zone of polyphosphate-free cytoplasmic material, located in nuclear area. RK fixation (2 hours), embedded in Epon, lead citrate post-stain. Scale marker 0.5 µ. — Reprinted from Voelz et al. (1966).

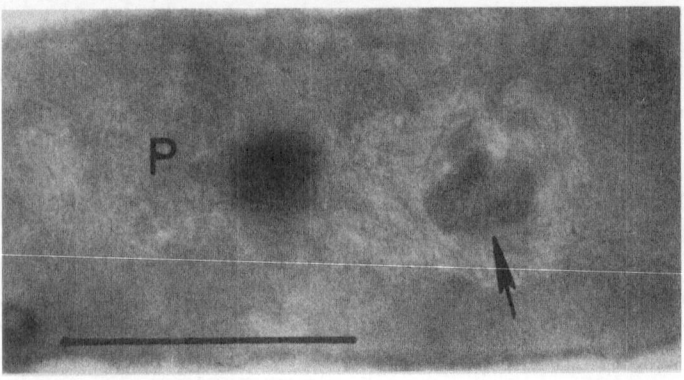

Fig. 70. *Bacillus subtilis*, core-like body in nuclear area (arrow). Fibrous state of DNA. P = Plasmalemmosome. Dark spot in center is artifact from uranyl post-staining. Scale marker 0.5 µ.

bodies in general on the intracellular distribution of nuclear material (see p. 64).

Polyphosphate granules are among the more frequent ones to be associated with the nuclear areas in bacteria and Cyanophyceae. Their formation was studied in detail by Voelz (1966) (Fig. 69). Other nuclear cores described in the literature (e.g. Fig. 70) may likewise represent polyphosphate bodies, but different fixation techniques used by different authors make a positive identification difficult.

Achromatic nuclear "cores" of unidentified nature were described by Murray (1953) and by Whitfield and Murray (1954). Fitz-James (1954, 1955 b) described ring-shaped nucleoids in early germination phases of *Bacillus* spores. The "cores" were followed microscopically through various phases of a procedure developed for the isolation of nuclear bodies and it was concluded, that the cores do not simply represent accidentally included portions of cytoplasm (Spiegelman et al. 1958). Also the fairly constant protein and RNA content of the nuclear fraction appears significant in connection with the possible existence of such non-DNA core areas (see Fitz-James 1958 and Butler and Godson 1964 for more details).

The opposite picture, a nucleoid surrounded by non-DNA basophilic material (RNA) was proposed by Hoffman (1951). The significance of this finding is questionable, since acid hydrolysis was used for the demonstration of DNA which tends to reduce the diameter of the nuclear body which then appears smaller than the basophilic area which can be stained with a solution of any basic dye (p. 62). More recently, however, uranyl-stained ultrathin sections were presented in which cytoplasmic areas rich in messenger RNA stand out in a darker tone. Such areas are preferably arranged around the nuclear bodies (Giesbrecht 1961, Kawata 1961, Fuhs 1965 b) (Fig. 48). In some instances electron staining is so intense that the ribosomes stand out as less deeply stained granules.

Nuclear Structures of Cyanophyceae

A. Introduction

Like in bacteria, the nuclear material of the Cyanophyceae is not organized as a true nucleus. Typical chromosomes, a nuclear membrane, and other accessory structures that characterize a true nucleus are absent. No spindle or other mitotic apparatus is found. Colchicine is without effect on division (Jakob 1950, Herbst 1952). As with bacteria, the only feasible way of demonstrating nuclear material is the demonstration of DNA. Therefore all nuclear "structures" referred to in this review, strictly spoken, are "regions characterized by the presence of DNA (DNA-plasm)". For a definition of the latter term see Kellenberger (1962), Fuhs (1965 d) or the introduction to this treatise.

Where such areas exhibit a certain regularity in number, shape, and arrangement, the idea of units or subunits at the microscopic level becomes suggestive, and the term "structure" may assume a more precise meaning. Investigation at the electron-microscopic level, however, shows that (as in bacteria) the only structural "backbones" of such formations are molecular fibres of DNA oriented roughly parallel to each other in a rather inconspicuous pattern. It also becomes apparent that the nuclear elements, if any, will lose their apparent individuality already by approaching each other sufficiently to establish contact, and that the nuclear patterns in general may be quite flexible.

In dealing with this situation it was felt desirable not to adhere to a rigid terminology but to use descriptive terms wherever possible. The literature is abundant in terms that are obsolete or of very limited value. Such, if at all, will be presented in quotation marks. Terms that are considered correct and useful according to present knowledge are introduced in s p a c e d typesetting.

B. Centroplasm and Nuclear Material

From the early days of cytology of the blue-green algae it has been known that the cells of most photosynthetic species contain pigmented and non-pigmented regions (Schmitz 1879). Strasburger in 1882 reportedly introduced the term c h r o m a t o p l a s m   for areas characterized by the presence of photosynthetic pigments and the term c e n t r o p l a s m for areas devoid of pigmented material. The latter term originally referred to the most conspicuous case of a centrally located colorless area surrounded by a peripheral pigmented layer.

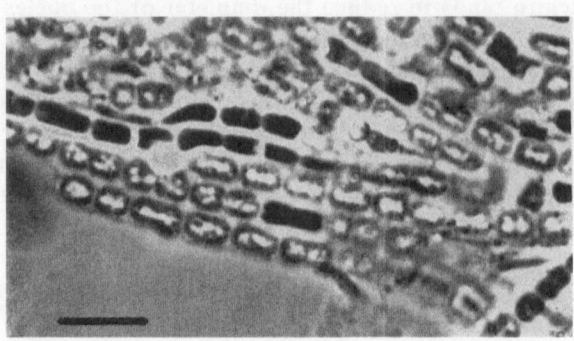

Fig. 71. *Anabaena cylindrica*, phase-contrast micrograph. Mounting medium: 30% serum albumin solution. Nuclear material is concentrated in a single central region which stands out as an area of low refractility. Scale marker: 10 μ.

Fig. 72. *Schizothrix mülleri*, Perenyi fixation, iron hematoxylin stain. Chromatin apparatus partly displaced by vacuoles. Scale marker 10 μ. — Reproduced from Guilliermond (1926).

Schmitz (1879) who worked in Strasburger's laboratory apparently was the first to associate the existence of the centroplasm with the idea of a nucleus, especially after basophilic substances were shown to be present in that area (Bütschli 1890).

In recent years the analysis of ultrathin sections has shown that regions not containing photosynthetic membranes are largely occupied by DNA-plasm, the remaining space consisting mainly of accumulations of ribosomes (Hopwood and Glauert 1960 c, Ris and Singh 1961, Fuhs 1963 b). In some species such as *Pseudanabaena catenata* Lauterb. and *Anacystis nidulans* the DNA-plasm occupies a single central region of the cell (Fig. 71) and the use of a term such as centroplasm or "central body" (Bütschli) appears understandable. In other species, however, the nuclear material is not arranged in the center or along the central axis of the cell, or such arrangement is a temporary one depending on the physiological state of the cell and the presence or absence of other cell inclusions (GUILLIERMOND 1926) (Fig. 72). Here the term "central body" appears inappropriate. This holds also for species that regularly show several separate portions of nuclear material per cell (see below).

C. The Morphology of the Nuclear Bodies, Earlier Findings and Methods of Demonstration

Early observers already found a predominance of linear elements in the nuclear structures of the blue-green algae. The appearance of the nuclear material after staining with basic dyes was described as a network of chromatic fibres by BÜTSCHLI (1890, 1896, 1900), HIERONYMUS (1892), NADSON (1895), HEGLER (1901), WAGER (1901), GUILLIERMOND (1906, 1925, 1926, 1933), DEHORNE (1920), HAUPT (1923) and LEE (1927).

Because this pattern resembles certain basophilic cytoplasmic elements in protozoa ("chromidia", now: endoplasmic reticulum) and because a limiting membrane is absent, the nuclear fibres were considered homologous with the chromidia. This led to the adoption of the term "chromidial apparatus" or "chromidial nucleus" by HERTWIG (1902) and GOLDSCHMIDT (1905). Later it was found that the nuclear structures of the Cyanophyceae are Feulgen-positive (POLJANSKY and PETRUSCHEWSKY 1929, NEUMANN 1930, DELA-PORTE 1940) whereas the true chromidia are not—they contain ribonucleic acid only. SPEARING (1937) therefore changed the former term into c h r o m a t i n a p p a r a t u s. The results with the Feulgen nuclear reaction were later confirmed in experiments with deoxyribonuclease (NEUGNOT 1950, FUHS 1958).

The difficulties in the cytological and cytochemical demonstration of nuclear structures in Cyanophyceae closely resemble those encountered in work with bacteria. For general information the reader is referred to Section II, C on p. 56. In Cyanophyceae as in bacteria, the Feulgen reaction and its modifications were the only specific staining reactions for DNA over a period of twenty years and, if applied correctly, were the only reliable techniques for the differentiation between nuclear structures and polyphosphate granules which latter are often referred to as "metachromatin". Unfortunately the Feulgen and related techniques do not represent the ultimate in preservation of fine fibrous structures. Also in the Cyanophyceae, acid hydrolysis produces a coarser pattern which results from the rounding up and disruption of fibrous elements (FUHS 1958). In this way many of the granular-reticular patterns described from Feulgen-stained or HCl-Giemsa preparations (e.g. CASSEL and HUTCHINSON 1954) are not free from artifacts, and it appears now that some of the most realistic pictures were obtained by those earlier authors who applied the iron-hematoxylin technique with enough care to stain DNA material only (HEGLER 1901, GUILLIERMOND 1906, 1926, HAUPT 1923, SPEARING 1937).

If some elementary care is exercised, useful information is also obtained by staining living Cyanophyceae with fluorescent dyes such as acridine orange or auramin. These dyes resemble other basic dyes in that both nuclear structures and polyphosphate granules are stained (von Zastrow 1953, Fuнs 1958). The second author found that another fluorescent dye, berberine sulfate, stained nuclear structures in a brilliant yellowish green without staining the polyphosphate granules (Fig. 79 g). This behavior is unique among the basic dyes and is found only with freshly prepared solutions of the dye. Upon prolonged standing, berberine sulfate solutions do no longer stain the nuclear structures but bind strongly to the poly-phosphate granules. This explains earlier different results by Krieg (1954 e). Widely encountered difficulties in reproducing results by others with acridine orange are due to a different behavior of different brands and batches of the dye. For details the reader is referred to Piekarski and Janssen (1967).

D. Structure of the Chromatin Apparatus, Light Microscopic Aspects

The idea of a true chromatic network was challenged by several workers who described individual threadlike or rodlike elements and sometimes called them chromosomes (Kohl 1903, Phillips 1904, Olive 1905, Guillier-mond 1925, Lee 1927, Neugnot 1950, Krieg 1954 e, Cassel and Hutchinson 1954). Guilliermond (1926) found the chromatin apparatus of *Phormidium favosum* and *Schizothrix mülleri* to consist of a number of threadlike elements, preferentially oriented parallel to the main axis of the trichome (Fig. 73). Neither the shape nor the arrangement of these elements did significantly change during the cell cycle. Guilliermond assumed that these threads divided by transverse fission or by being dissected by the newly formed cross-wall. He realized that this mode of division, although different from mitosis, was equally effective in distributing complete sets of genetic information ("Dénomitose", Chatton and Poisson 1930, "Haplomitose", Guilliermond 1933).

Spearing (1937) among other Cyanophyceae investigated an unidentified *Oscillatoria* with 3 μ wide trichomes. He found a chromatin apparatus that consisted of several individual rod-shaped elements which were oriented parallel to the main axis of the trichome (Fig. 73), and the number of which per cell was not constant but varied from five to seven, thus distinguishing the chromatin apparatus from an ordinary set of chromosomes. This author also was successful in recognizing rod-shaped nuclear elements in living, unstained cells of *Phormidium retzii* (Spearing 1961).

Information confirming and extending Spearing's findings was obtained with *Oscillatoria amoena* (Kütz.) Gom. (now: *Microcoleus vaginatus*) (Fuнs 1958). In this investigation the ribonuclease-pepsin-Giemsa technique was used which combines specificity and gentle action (Boivin et al. 1947 a, Peters and Wigand 1953, see also above, Section II, C, p. 63). In this procedure, polyphosphate granules are largely removed by the acid buffer used with the pepsin. The nuclear material appeared in a reddish violet color, and superposition of linear elements could be easily recognized as an increase in optical density at the points of intersection or by scanning with the fine focus control. Unnecessary shrinking of the specimens was avoided by using as mounting media either water or dilute aqueous solutions.

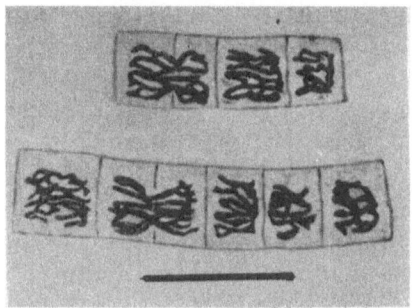

Fig. 73 *a*. *Phormidium favosum*, Perenyi fixation, iron hematoxylin stain. Division stages of chromatin apparatus. Scale marker: 10 µ. — From GUILLIERMOND (1926).

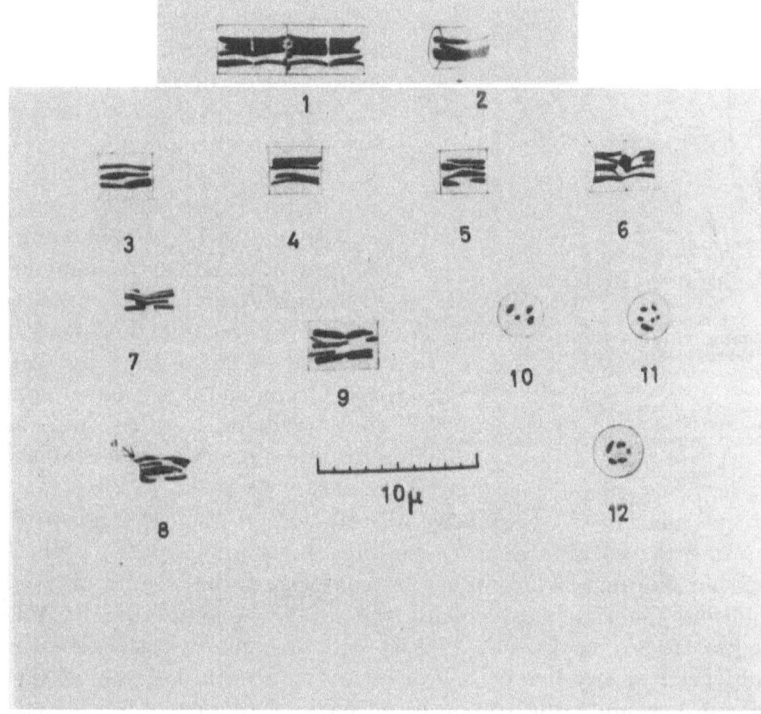

Fig. 73 *b*. Unidentified species of *Oscillatoria*. Belling's fixation. — 1. A rather condensed drawing of two dividing cells showing the transverse breaking of the longitudinal chromatin strands, iron hematoxylin stain. — 2 through 12. Iodine — gentian violet stain. — 2. A perspective view of a fully grown cell showing one clearly distinct chromatin rod, the others being somewhat clumped together. — 10 through 12. End view in sections (photographic style of drawing). There are six or seven rods visible in the cross section, and some are thicker than others. — Reprinted from SPEARING (1937).

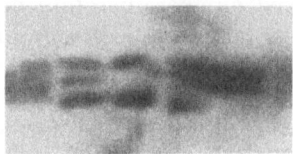

Fig. 73 *c*. Photograph of cell shown in Fig. 73 *b*, part 9. — Reprinted from original negative which was kindly provided by the author. — From SPEARING (1937).

This alga resembles SPEARING's *Oscillatoria* in containing a number of rod-shaped nuclear elements per cell which are oriented parallel to each other and in the main axis of the trichome. Besides this, the species exhibits some other properties which permit a better insight into the significance and mode of division of the nuclear elements.

In a rapidly growing trichome of a filamentous blue-green alga most cells are in some stage of division. The cells that just have originated by cell division are the shortest. Immediately after formation they start growing longer and the next cross-wall begins to appear in the form of an iris-like diaphragm. As soon as the cells have attained twice their minimum length, the cross-wall is completed and a new cycle initiated. The distribution of cell lengths (Fig. 74) indicates that cell growth proceeds approximately linear with time. The uninterrupted sequence of division cycles resembles that found in bacteria during unrestricted growth, and the overall pattern of nuclear replication appears to be similar in this respect. GEITLER (1960, p. 46) states: "Die Erscheinung der zeitlich einander nahegerückten oder fast zusammenfallenden Teilungen führt die primitive Organisation der Blaualgen anschaulich vor Augen. Bei kernführenden Organismen ist ein solcher Vorgang nicht möglich..." As to absence of "resting stages", similar statements were made by DEHORNE (1920).

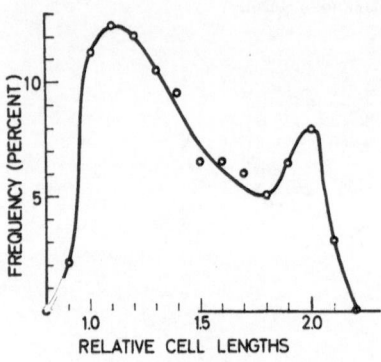

Fig. 74. Frequency distribution of cell lengths in filaments of *Microcoleus vaginatus* (*O. amoena*), based on measurements of 620 cells of a growing population. The curve suggests exponential increase in cell lengths throughout the cell cycle with a possible delay prior to division. For similarity with distribution of age classes in growing bacterial populations see POWELL (1956).

In *O. amoena* (*M. vaginatus*) all rod-shaped nuclear elements are of the same size, their length almost equalling the minimum cell length. Contrary to earlier findings which must be considered the results of poor optical resolution, the rod-like elements do not grow longitudinally and do not undergo transverse fission. During cell growth, two sets of elements can be observed in any one cell, each set moving with one pole of the cell and each one becoming the chromatic apparatus of one of the daughter cells. In cells which are approximately halfway between divisions, the overlap of the two sets of elements in the central portion of the cell is easily seen (Fig. 75). Single elements can best be observed in the peripheral layers of the cell (Fig. 76). In this alga, each set comprises about four elements, the most frequently observed number therefore being eight per cell. Since the observations also suggest longitudinal fission of the elements as a mode of replication, the following scheme of the cell cycle appears to be in accordance with the known facts (Fig. 77). In considering longitudinal fission as a mode of replication, it should be considered, however, that the chromatin elements are not to be compared with chromosomes but closely resemble bacterial nucleoids (see below, p. 146).

Other observations on *O. amoena* (*M. vaginatus*) support the idea that the nuclear elements represent complete sets of genetic information, each element representing one genome. The ends of the trichomes in this species normally consist of a tapered zone which is five or six cells long. The

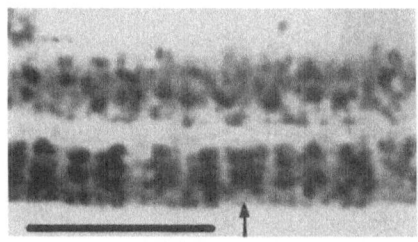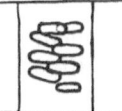

Fig. 75. *Microcoleus vaginatus* (*O. amoena*), ribonuclease Giemsa technique, nuclear material and some polyphosphate granules. The cell marked by an arrow is shown diagrammatically. The diagram is derived from visual analysis which in contrast to the photograph includes all optical planes of the cell. The provisional sketch was redrawn in India ink on an enlarged print under careful analysis of the density of the silver grain. Subsequently the silver grain was removed by chemical bleaching of the print and the drawing photographically reproduced and reduced in size. — The nuclear material forms rod-like elements of approximately equal size. Scale marker 10 μ. — Reprinted from FUHS (1958).

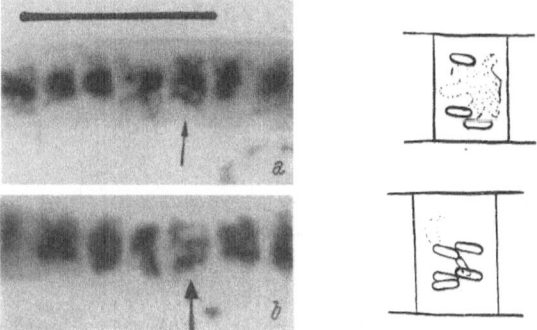

Fig. 76 *a, b. Microcoleus vaginatus* (*O. amoena*), Chabaud's fixative, ribonuclease, pepsin, Giemsa stain. Chromatin apparatus. See diagram for interpretation and legend to Fig. 75 for details of preparation of diagram. Scale marker 10 μ. — Reprinted from FUHS (1958).

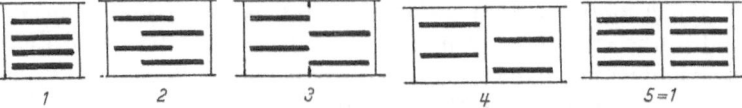

Fig. 77. Schematic representation of nuclear replication and segregation of chromatin elements in polyenergidic cells of *M. vaginatus* (*O. amoena*). — Reprinted from FUHS (1958).

entire formation resembles a bottleneck except for the enlarged and almost spherical terminal cell (Fig. 78). These tapered ends are formed by hormogonia and broken filaments upon resumption of growth in a very simple manner: Viable end cells of normal diameter undergo cycles of growth and division whereby the diameter of the cell at the distal side is increasingly reduced. Finally, the terminal cell assumes its characteristic spherical shape.

Cytological investigations show that the spherical end cells no longer divide. The cells of the "bottleneck" portion, however, divide actively, but the number of nuclear elements per cell is reduced. The smallest cells adjacent to the terminal cell contain one chromatin element only, which undergoes replication. The two daughter elements move in opposite directions followed by cell division. In more proximal regions of the "bottleneck" a similar pattern can be observed with two to four or three to six nuclear elements involved (Fig. 79).

One may conclude from these observations that (1) one nuclear element is capable of furnishing the genetic information needed for survival and replication of complete cells, and that (2) a cell containing one single element is capable of producing progeny comprising cells with more than

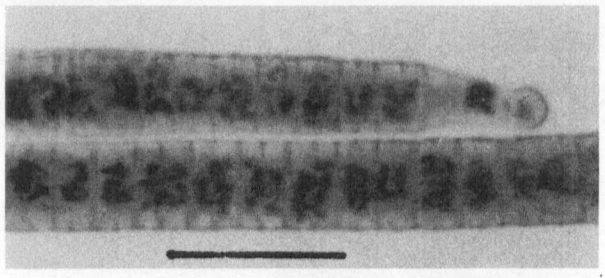

Fig. 78. *Microcoleus vaginatus* (*O. amoena*). Neutral red stain (non-specific). The figure shows the shape and cellular structure of a fully developed trichome tip consisting of a bottleneck region and a non-dividing, non-pigmented end cell. The scale marker represents 10 μ. — Reprinted from Fuhs (1967).

one element and even cells with a normal set of four to eight elements. (This is evident from the fact that despite cell divisions the bottleneck remains four to five cells long.) The four to eight elements then must be replicas of the single element and therefore must be identical. Thus in *O. amoena* (*M. vaginatus*), one chromatin element represents a complete genome, and cells with more than one element are p o l y e n e r g i d i c (for definition see p. 4).

To explain the observed formation of monoenergidic end cells one may advance the hypothesis that rupture of the filament and exposure of a cross-wall to the environment interferes with nuclear divisions (Fuhs 1958).

A developmental cycle involving monoenergidic and polyenergidic phases is found among members of the order *Chamaesiphonales*. Beck (1963) investigated a strain of *Pleurocapsa fuliginosa* which starts out with a spore containing a single chromatin element. Upon germination the organism grows into a vesicular cell about 50 μ large which contains a large number of identical chromatin elements. Subsequently each chromatin element becomes surrounded by a portion of cytoplasmic material and a membrane. In this manner the entire cell is converted into a mass of endospores which become liberated by rupture of the membrane of the mother organism. This recent investigation also provides a clue to early observations on the related form, *Dermocarpa fucicola* (Gardner 1906) (Fig. 80). The chromatic network described by Gardner very probably constitutes

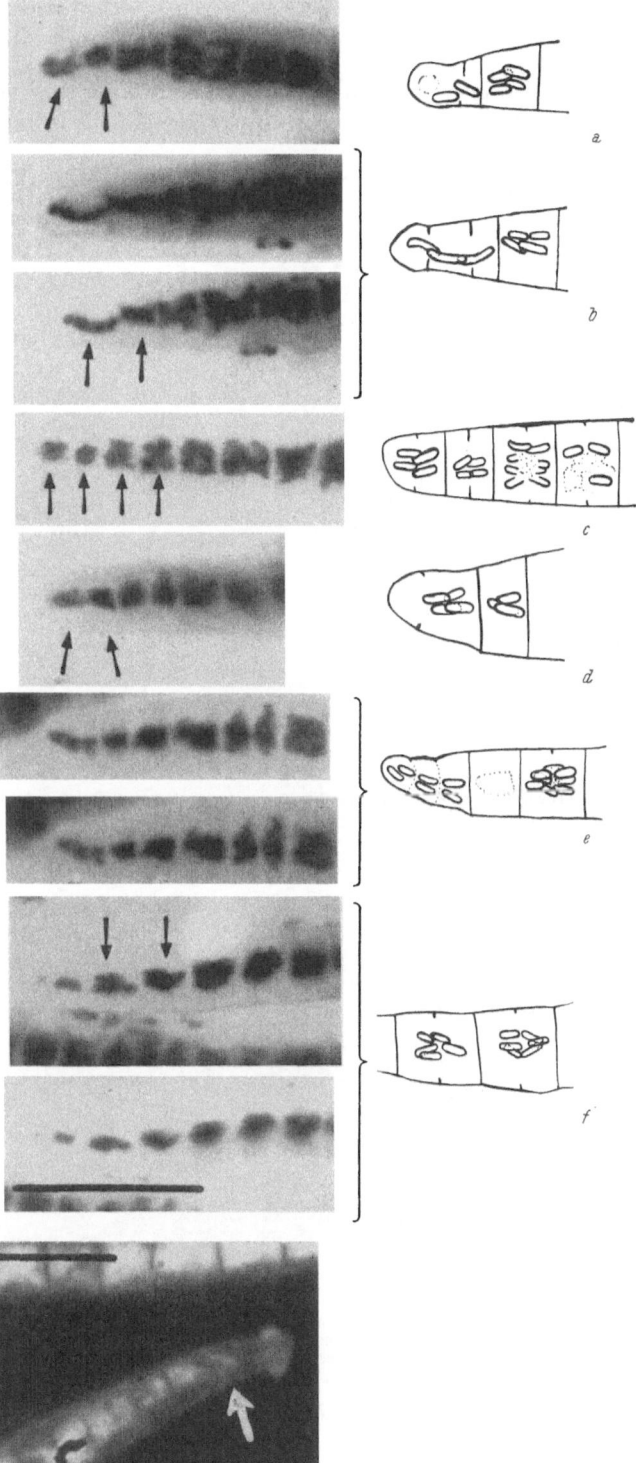

Fig. 79. *Microcoleus vaginatus (O. amoena).* — *a* through *f.* Chabaud's fixative, ribonuclease, pepsin, Giemsa stain. For interpretation see diagrams, legend to Fig. 75, and text. — *g* Berberin sulfate vital stain, fluorescent microscopy. Scale markers represent 10 μ. — Reprinted from FUHS (1958).

a polyenergidic protocaryon which during endospore formation disintegrates into chromatic elements representing one genetic complement each.

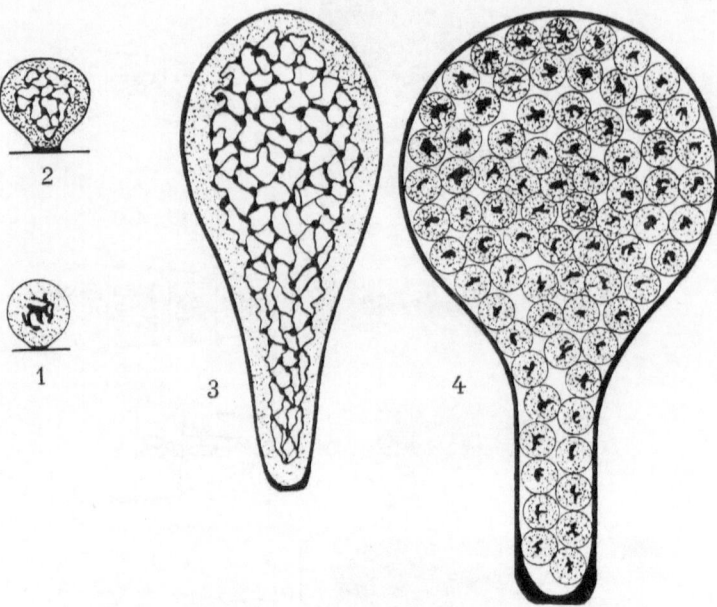

Fig. 80. Developmental cycle of *Dermocarpa fucicola*. Nuclear material shown. — Reprinted from GARDNER (1906).

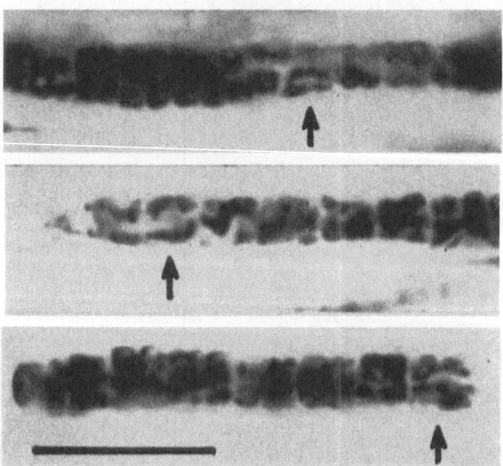

Fig. 81. *Oscillatoria tenuis*. Chabaud's fixative, ribonuclease, pepsin, Giemsa stain. Chromatic apparatus in some cells shows separation into two, sometimes unequal masses. Scale marker represents 10 μ.

More generally, the following characters may be taken as indication of polyenergidy in Cyanophyceae: (a) the appearance of several distinct nuclear elements per cell (*O. amoena, Nostoc spec.*, Fuhs 1958), (b) the appearance of reduced amounts of chromatin in healthily dividing end cells (*O. amoena, O. tenuis*, Fuhs 1967), and (c) the occasional appearance

of a chromatic network which is split into two or more masses, sometimes
of unequal size (*O. prolifica,* Fuhs 1967) (Fig. 81).

An event of uncertain significance is the germination of a k i n e t e s
(resting cells) which frequently are much larger than vegetative cells and

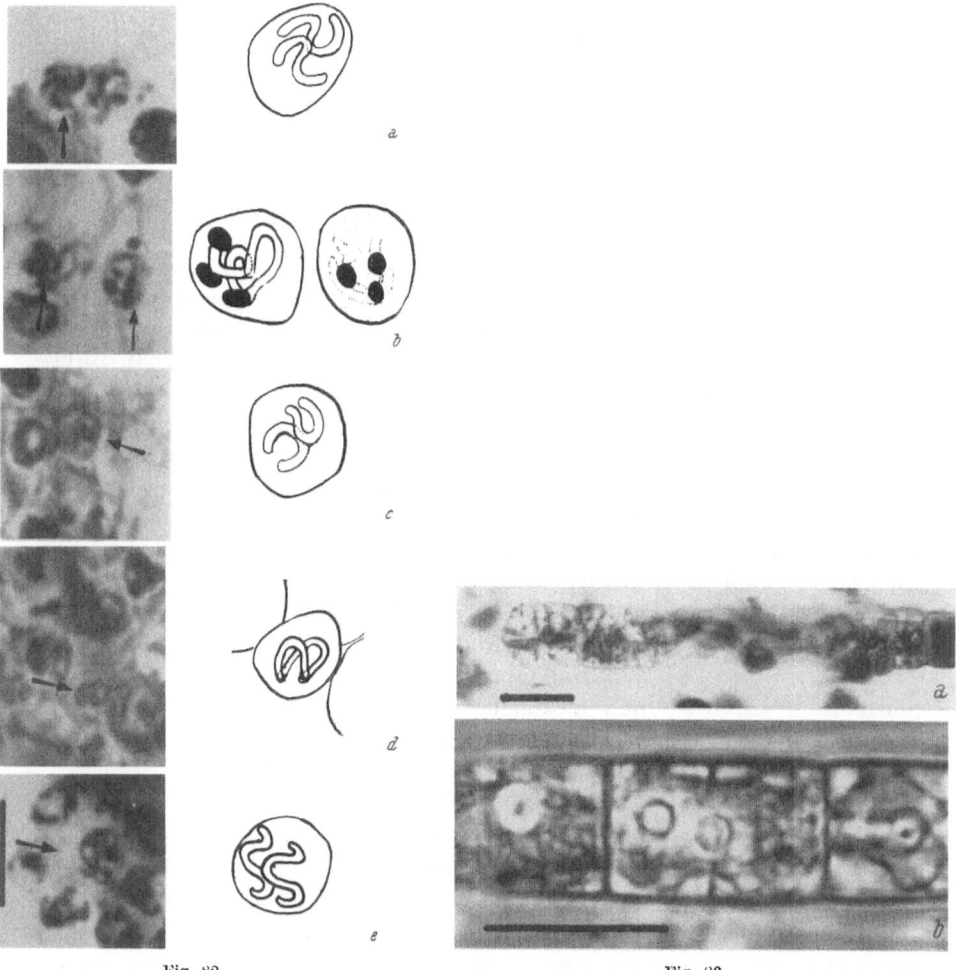

Fig. 82.   Fig. 83.

Fig. 82. *Gloeothece* spec., Chabaud's fixative, ribonuclease, pepsin, Giemsa stain. Chromatic apparatus consisting
of two equal chromatin elements. See diagrams and legend to Fig. 75 for interpretation. In *b* some polyphosphate
granules. Scale marker represents 5 μ. — Reprinted from Fuhs (1958).

Fig. 83. *Oscillatoria borneti.* — *a* Chabaud fixation, ribonuclease, pepsin, Giemsa stain. Nuclear material in central
part of the cells. — *b* Live filament mounted in 10% serum albumin solution, phase-contrast. Polyphosphate
granules and keritomy of cytoplasm. Scale markers represent 10 μ.

contain a voluminous sponge-like chromatin apparatus (Poljanski and
Petruschewski 1929). At germination, divisions may follow each other
rapidly and without indication of growth (Geitler 1936, 1942, 1960).

Equally uncertain is the caryological status of the whip-like parts of
the trichomes of the *Rivulariaceae.* The cells in these regions are vacuolized
which is considered to indicate reduced viability (Geitler loc. cit.).

The appearance of n a n n o c y t e s (very small cells) has been described in the *Chroococcales* and in a similar form in the *Nostocaceae* (GEITLER 1960, LAZAROFF and VISHNIAC 1962). Nannocytes appear to be derived from cells of regular size by a sequence of divisions not accompanied by growth. Determinations of DNA content would provide clues as to the significance of these processes.

A representative of the *Chroococcales* resembling *Gloeothece* was found to contain two identical chromatin elements per cell which obviously were ready for distribution into daughter cells (Fig. 82). Cell division stages were absent in this population of cells from a slowly growing or stationary crude culture. This indicates that duplication of the chromatin follows rather than precedes cell division, a process similarly indicated in *O. amoena*.

Reports on granular chromatin elements in Cyanophyceae are few and probably due to inadequate methodology. FUHS (1958) occasionally found massive coalescence of nuclear elements into rounded bodies as a result of acid hydrolysis. The contours of these aggregates are somewhat diffuse. Descriptions of sharply defined granules indicate confusion with polyphosphate granules. Erroneous descriptions of this sort are common throughout the literature and are occasionally found in descriptions of notoriously critical workers. In this manner, FUHS (1967) was unable to confirm DRAWERT's (1949/50) observation of granular chromatin elements in *Oscillatoria borneti*. Treatment with Chabaud-ribonuclease-pepsin-Giemsa revealed a centrally located DNA-containing area (Fig. 83 a). DRAWERT's Feulgen-positive granules probably were undigested residues of polyphosphate granules which in this species can be extraordinarily large $(3\,\mu)$ (Fig. 83 b) and undoubtedly will resist acid hydrolysis to a considerable extent. Upon treatment with leucofuchsin they will appear in a much more brillant red color than the DNA.

E. Electron-Microscopic Evidence

When bacteria were first investigated by the ultrathin sectioning technique, solid strands were observed in the nuclear areas, sometimes resembling rows of interconnected beads. With similar techniques, identical pictures were obtained with *Oscillatoria princeps* and *O. amoena* (SHINKE and UEDA 1956, HAGEDORN 1960, 1961). As in bacteria, the solid strands were surrounded by empty spaces and therefore can be considered identical with bacterial nuclear bodies in their "artificially condensed state" as defined earlier (p. 89). Fig. 84 illustrates this type of nuclear artifact in *Pseudanabaena catenata* Lauterb.

Kellenberger's fixation (RYTER and KELLENBERGER 1958) was successfully applied to several blue-green algae. The results resemble those obtained with bacteria. The nuclear areas display the "homogenous state" characterized by very fine fibres of DNA in a somewhat irregular arrangement (HOPWOOD and GLAUERT 1960 c, RIS and SINGH 1961, ROSSNER 1963, PANKRATZ and BOWEN 1963, ECHLIN 1964, LEAK 1964, 1965, LEAK and WILSON 1965) (Fig. 85). ECHLIN (1962) obtained quite similar results with glutaraldehyde

and osmium fixation (no tryptone added) followed by uranyl acetate treatment (for an assessment of this technique see p. 98). A correct estimate of
the number of nuclear elements per cell is principally impossible for the
same reasons as in bacteria. Absence of a nuclear membrane prohibits the
exact differentiation of nuclear elements which can coalesce to form a
coherent DNA-plasm. Large forms in which light-microscopic observations
indicate the existence of separate nuclear entities would require extensive
serial sectioning, a technique which with objects as large as these presents
almost insurmountable difficulties.

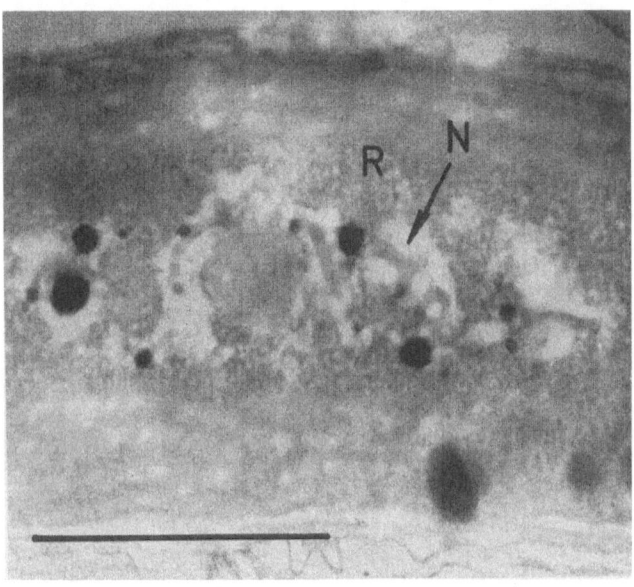

Fig. 84. *Pseudanabaena catenata*. Osmium-dichromate fixation, no uranyl treatment, Vestopal-embedded, thin
section. Cell shows "centroplasm" composed of ribosomes (*R*) and artificially condensed strands of nuclear
material (*N*). Scale marker 0.5 µ.

Attempts to produce the "fibrous state" of the DNA-plasm as defined
earlier were successful (Fig. 86), but a more detailed analysis of nuclear
fine structure in serial sections of a smaller form is still lacking. Nevertheless all results presented thus far clearly place the
structure which we called a "chromatin element" in
close relationship to the bacterial nucleoid. The compact chromatin apparatus of the smaller Cyanophyceae in every respect is indistinguishable from
a bacterial nucleoid.

F. Cytochemical Evidence

As mentioned above, the presence of DNA in Cyanophyceae was established with the aid of Feulgen's nuclear reaction and with nucleases. Biochemical studies on the DNA of Cyanophyceae are extremely scarce.
MOCKERIDGE (1926) demonstrated the constituents of ribonucleic acid in

Nostoc: Biswas (1956) found deoxyribose (Dische reaction). Biswas and Myers (1960) determined the DNA content of *Anacystis nidulans* as 0.75 to 0.94 per cent of the wet weight and demonstrated the presence of

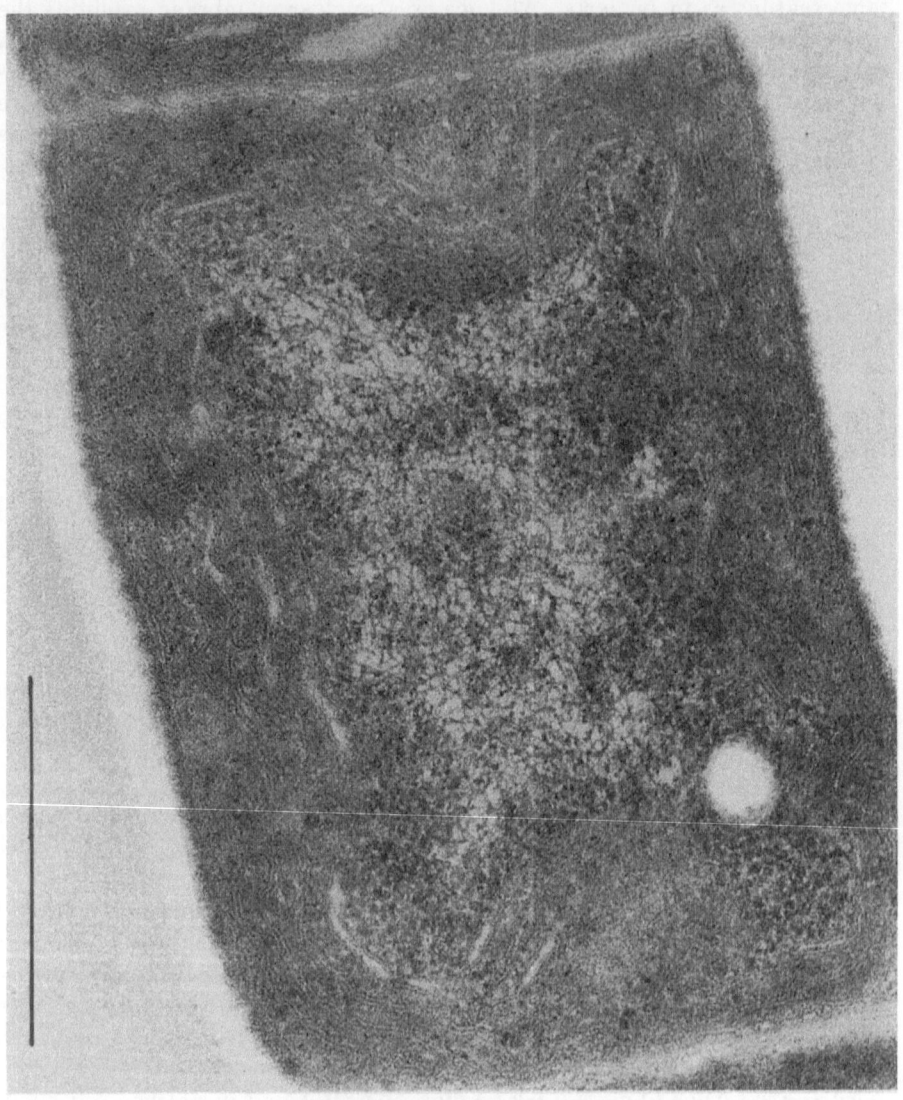

Fig. 85. *Phormidium calothricoides*, RK fixation. Homogeneous state of DNA plasm. Scale marker 0.5 μ. — Courtesy Dr. M. R. Edwards, Division of Laboratories and Research, New York State Department of Health, Albany.

guanine, adenine, cytosine and thymine in molar ratios of 25.5 : 23.3 : 24.8 : 22.3. The A/T ratio is almost unity, but the G/C ratio is significantly different (1.19). Edelman et al. (1967) found GC (guanine + cytosine) contents in a number of Cyanophyceae to range from 38 to 56 per cent; Kaye et al. (1967) in *Plectonema boryanum* found 47 per cent.

LEAK (1965) used tritiated thymidine to label the DNA of an *Ana-baena* sp. and with the aid of autoradiography demonstrated its incorporation into the chromatic apparatus.

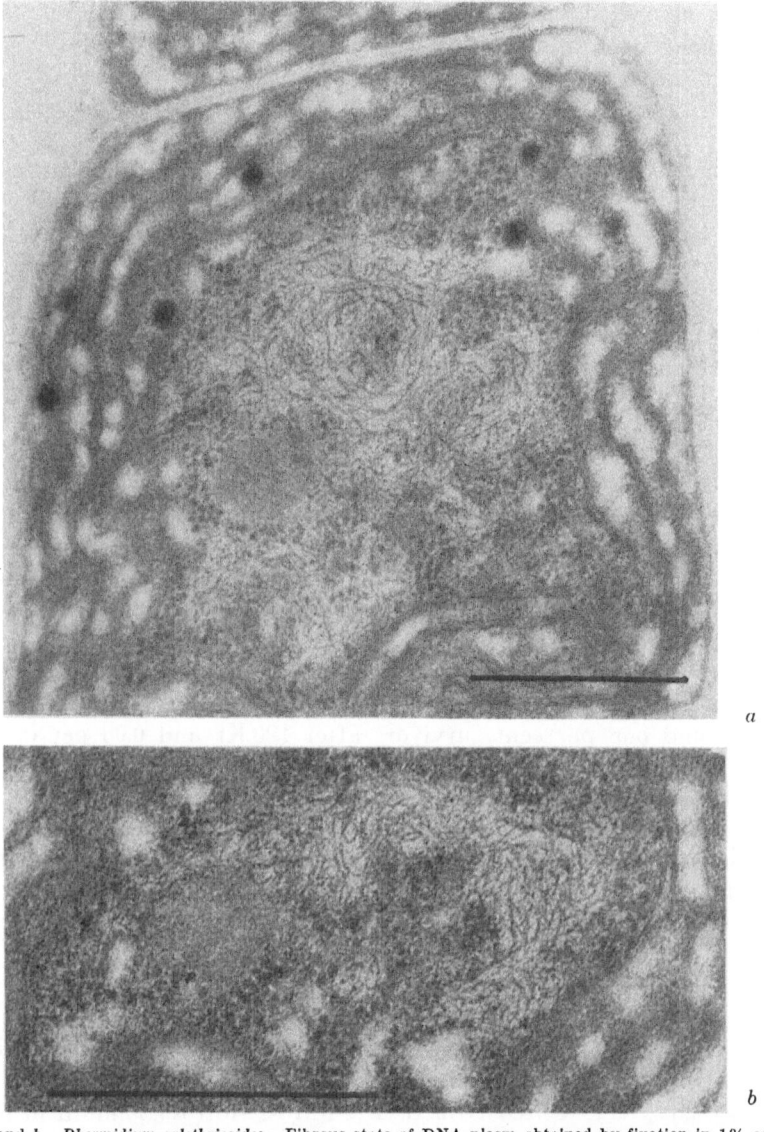

a

b

Fig. 86 *a* and *b*. *Phormidium calothricoides.* Fibrous state of DNA plasm obtained by fixation in 1% osmium tetroxide in veronal acetate buffer at 5 to 10° C for 3 hours, uranyl acetate 0.5% in buffer 40 minutes, dehydration in ethanol and embedding in Epon. Scale markers 0.5 μ. — Courtesy Dr. M. R. EDWARDS, Division of Laboratories and Research, New York State Department of Health, Albany.

Basic amino acids were found to be present in both the chromatin apparatus and the surrounding cytoplasm, but the association of any of these with DNA is uncertain (BISWAS 1957). This holds also for the protein component of a preparation described as "deoxyribonucleoprotein" (BISWAS

1961). Electron microscopic evidence clearly indicates that the DNA of the Cyanophyceae is not associated with histone (cf. p. 51). De and Ghosh (1965) presented histochemical evidence for the absence of histone in several Cyanophyceae; Biswas (1957) found the chromatin apparatus to be more resistant against trypsin and pepsin than chromosomes are. (Trypsin preparations containing residual nuclease activity will also attack the chromatin apparatus.)

G. X-Ray and Ultraviolet Resistance and Mutants

Herbst (1952) determined the X-ray resistance of an unidentified *Oscillatoria* species the trichomes of which were 5.4 μ wide. The cells were 2.7 to 3.6 μ long. The chromatin apparatus consisted of several bar-like elements oriented parallel to the main axis of the trichome. Irradiation with soft X-rays was tolerated up to 700 Kr without any apparent negative effect on growth response or cell structure. Application of 1,050 Kr proved to be lethal for all cells. At a dose of 900 Kr the trichomes contained dead, damaged, and healthy cells. The diameter of the cells which grew out of the damaged ones was significantly reduced. Some stable variants were also obtained which were characterized by a reduced cell width (one third to one half of the normal) and by altered pigment content. With hard X-rays, complete tolerance was observed up to 200 Kr whereas 900 Kr represented a lethal dose. For comparison, the same experimental arrangement was used for the irradiation of different organisms. *Chlamydomonas* was killed by 100 Kr of soft rays and miscellaneous bacteria at about 200 to 300 Kr. This places the *Oscillatoria* used in these experiments in the rank of the most X-ray resistant organism known.

Anacystis nidulans was found to be considerably less resistant. Moore (1966) found one per cent survivors after 120 Kr and 0.08 per cent after 180 Kr. Kumar (1964) reports comparable results.

These results seem to indicate that X-ray resistance is a function of polyenergidy. The extremely resistant *Oscillatoria* displayed several rod-like nuclear elements per cell and therefore must be considered polyenergidic, whereas the small, bacteria-like *Anacystis* may contain one or two energids per cell. The appearance of trichome segments of reduced diameter during recovery of the *Oscillatoria* may indicate the elimination by X-rays of some of the chromatic elements.

Singh and Singh (1964) and Kumar (1966) used ultraviolet light to obtain mutants of Cyanophyceae. Moore (1966) found that *Anacystis nidulans* was less sensitive against ultraviolet than *Escherichia coli* strain B/r.

Van Baalen (1965) used N-methyl-N-nitro-N-nitrosoguanidin to produce mutants of *Anacystis nidulans*.

H. Genetic Exchange

Kumar (1962) reports on possible genetic recombination in *Anacystis nidulans*. Stable penicillin and streptomycin-resistant strains were obtained by gradual adaptation to these antibiotics. Upon mixing of inocula of the two strains in new medium and various periods of growth, double resistant isolates were obtained in some of the subcultures. These experiments resemble early studies on genetic recombination in *E. coli* K 12 except that with *Anacystis* an inoculum of more than one cell is usually required to obtain growth. The rate of recombination was tentatively determined at 10^{-8} resembling that of low frequency mating in *E. coli* K 12. SINGH and SINHA (1965) report on genetic recombination in *Cylindrospermum maius* cultures.

KUMAR et al. (1967) found that a strain of *Anacystis nidulans* was rendered more sensitive against streptomycin by repeated subculture on a medium containing proflavine. This is taken to indicate the possible location on an episome-like element of the marker for streptomycin resistance. SINGH (1967) discusses the possible genetic control of sporulation (akinete formation) in *Anabaena doliolum*.

J. Obsolete Terms Allegedly Referring to Nuclear Elements

Occasional reappearance of obsolete concepts recommends that these be mentioned briefly.

"Endoplasts" are said to be non-essential components of the centroplasm characterized by moderate basophilia. They correspond to a negative replica of the chromatin apparatus. They may be tentatively correlated with islands of ribosomal material, but since these accumulations are not organized as discrete organelles, the use of this term should be discontinued. SPEARING (1961) mentions that dilute methylene blue solutions and similar stains can produce colored vacuoles which may give rise to similar erroneous interpretations.

The latter author in his otherwise very critical paper revives the concept of the "nucleosome". It should be noted, however, that one of the originators of this concept in her most recent article (HOLLANDE 1962) contradicts herself by identifying in one instance the nucleosome with a DNA granule associated with a (hypothetical) non-DNA carrier structure and in another instance with a non-DNA granule which most obviously is associated with the nucleoid or the chromatin apparatus respectively. The complete terminology used by this author is not reviewed here.

"Karyoids" described in Cyanophyceae by BRINGMANN (1950) are polyphosphate granules. The apparent misinterpretation is due to methodological problems discussed earlier (p. 58). The nuclear structures were apparently overlooked by this author.

K. Nuclear Structures in Non-Pigmented Cyanophyceae

The non-pigmented or apochlorotic Cyanophyceae will be considered separately, since they still are incompletely investigated organisms. There is some indication, however, that their organization does not materially differ from that of other Cyanophyceae.

BÜTSCHLI (1890, 1896) reports on a "central body" in *Beggiatoa*. GARDNER (1906) describes a similar element. A most comprehensive review of earlier findings was given by DELAPORTE (1939/40). PETTER (1933 b) used the Feulgen reaction and found only a diffuse staining of the cells of *Beggiatoa*. According to DELAPORTE, an axial strand of chromatin can be demonstrated in very young filaments only, while older cells show diffuse or granular staining. An axial filament was also found by BAHR and SCHWARTZ (1957) who investigated filaments of *Thiothrix* as well. DEVIDÉ (1949), VON ZASTROW (1953) and DRAWERT and METZNER-KÜSTER (1958) report on essentially negative findings. VON ZASTROW compares *Beggiatoa* cells with cells of Cyanophyceae from old cultures, with vacuoles and disseminated portions of DNA-plasm. Granular DNA material was also found by MAIER (1964). In ultrathin sections, areas of DNA-plasm were observed (MAIER and MURRAY 1964).

L. Symbiotic Cyanophyceae

Intracellular structures resembling a unicellular blue-green alga were found in flagellates and Chlorophyceae (GEITLER 1923, 1936). GEITLER (1959 a, b) could neither prove nor disprove the existence of nuclear material in these formations but found a central inclusion which was definitely non-nuclear. In ultrathin sections, HALL and CLAUS (1963) showed the existence of a fibrous material in a halo-like area around the central body and the binary fission of this halo-central body complex. The fibres resemble DNA.

Synthesis and Conclusions

All evidence currently at hand and presented in the preceding text is compatible with the assumption that bacterial nucleoids and the chromatin elements of blue-green algae are identical structures. The principal arguments pertinent to this point will be summarized and discussed briefly.

Size and Shape. Both nucleoids and chromatin elements are elongated structures of similar size. The smallest Cyanophyceae resemble in size the smaller bacteria, and their nuclear elements are indistinguishable from bacterial nucleoids in every respect. Chromatin elements in some Cyanophyceae are larger, but the difference is not very pronounced, since the largest chromatin elements thus far observed measure 0.23×3.5 microns (*Gloeothece spec.*, FUHS 1958). The large-celled forms in both bacteria and Cyanophyceae are polyenergidic, i.e. they contain more than one nucleoid or chromatin element per cell.

Both nucleoids and chromatin elements are not bounded by membranes that would separate them from the cytoplasm. If more than one genetic element is present in any one cell, the genetic material may coalesce into a network or even into a single undifferentiated mass of DNA-plasm that cannot be resolved with light- or electron-microscopic means. Although reports that the chromatic apparatus of the Cyanophyceae consists of single linear elements have become more frequent in recent times and have considerable aided in its analysis, these reports do not imply that all reticular structures are products of poor preparation or inaccurate observation. On the contrary, the existence of compound nuclear structures, reticular or other, is a serious possibility in many Cyanophyceae (ROSSNER 1963) and in larger bacteria, and will continue to interfere with the determination of the number of energids per cell in such species.

Fine Structure. In ultrathin sections, both nucleoids and chromatin elements show identical patterns consisting of fibres of DNA, not of deoxyribonucleohistone. The fibres are preferentially oriented parallel to the longest axis of the genetic element wherever such axis is apparent. In bacteria the parallelism of the fibres does not indicate polyteny in the conventional sense of chromosomal polyteny. The fibres represent the segments of a single genophor. Although attempts to isolate entire DNA molecules from Cyanophyceae have not been made, it would appear unreasonable to explain the parallelism of DNA fibres in the genetic elements of the Cyanophyceae in any other way, because this would leave the Cyanophyceae with a genetic map only $1/100$ to $1/1000$ as long as that of a bacterium. So small an amount of genetic information is most unlikely in view of their primarily autotrophic mode of life and their level of cellular organization which in many instances is higher than that of any bacterium.

Replication. In describing the light-microscopic appearance of the chromatic apparatus in Cyanophyceae it was noted that the chromatin elements quite often are arranged side by side and appear to segregate by gliding along each other. This suggests longitudinal fission of the elements as a possible mode of replication. While such concept appears quite satifactory in the light of the traditional postulate of longitudinal fission as the only feasible way of equal distribution of genetic information stored on a linear structure, the concept is much less satisfactory in the light of recent models of nucleoid structure. Here longitudinal fission takes place at the molecular level only, i.e. during replication of the genophor. On the cytological level, longitudinal fission of the nucleoid is difficult to visualize (Fig. 54) and most likely must be replaced as a model by a process that under certain conditions mimics longitudinal fission but in reality consists in the buildup of a new genetical element in the neighborhood of the existing one (Fig. 66). A similar situation may exist in the Cyanophyceae.

Membrane Attachment and Coordinated Behavior. Possible mechanisms of distribution into daughter cells of more than one genetic element have been considered for bacteria in view of the coexistence of nucleoids and episomal elements (p. 32, Fig. 11). In this model, membrane attachment in some form is considered essential for the orderly distribution of the different genomes.

In the polyenergidic, filamentous Cyanophyceae, the chromatin elements replicate and are distributed into daughter cells in equal or approximately equal numbers as shown in Fig. 77. Projection of Fig. 11 into Fig. 77, however, reveals an important discrepancy. The bacterial model (Fig. 11) which implies permanent attachment of the genetic elements will operate only if attachment occurs at the longitudinal segment of the wall and if the sites of attachment separate in a specific manner requiring a specific pattern of membrane or wall growth. Obviously the model does not apply to Cyanophyceae, since in these forms the chromatin elements often are quite remote from the longitudinal walls. This applies to both smaller and larger Cyanophyceae as evident from light microscopic observations and from the undisturbed arrangement of the photosynthetic lamellae along the periphery of small-celled species. Also the mode of multiple cross-wall formation seen in *Oscillatoria* and related forms is in conflict with the bacterial model, since it suggests diffuse rather than localized growth of the membrane-wall complex. It remains to be seen whether these difficulties reflect the inadequacy of present models or a basic difference in organization of both major groups of protocaryotic organisms.

A certain degree of divergence becomes apparent in the nuclear organization of polyenergidic bacteria and Cyanophyceae in that in bacteria the genomes are arranged in series (Fig. 18) while in Cyanophyceae they are oriented parallel to each other (Figs. 73 and 75).

The latter arrangement bears some resemblance to nuclear patterns in meso- or eucaryotic organisms which show a greater number of chromosomes that are oriented parallel to each other and segregate by moving alongside each other and perpendicular to the plane of division without the apparent involvement of a mitotic spindle (Heliozoa, GRELL 1953; amoebae, VAHLKAMPF 1905; Euglenales, DEHORNE 1920, TSCHENZOFF 1916; dinoflagellates, ENTZ 1921, CHATTON 1918, CHATTON and POISSON 1930, SKOCZYLAS 1954, DODGE 1960). These chromosomes also resemble the chromatin elements of Cyanophyceae in persisting throughout the division cycle. GRELL (1953) found that "Sammelchromosomen" in *Aulacantha scolymantha* represent a complete genome each and that the highly polyploid state of this organism does not require the distribution of exact numbers of chromosomes to the daughter cells. For a more detailed discussion of the phenomena of endomitotic polyploidization and genome segregation in protozoa, the reader is referred to GEITLER's (1953) contribution to this series. In a review (1942) this author also discusses in some detail the nucleus-like behavior of the chromatin apparatus in Cyanophyceae.

The latter type of genome segregation implies a more active involvement of the genetic material in the segregation process proper than does the quasi-mitotic process suggested for bacteria by JACOB et al. (1963) (Fig. 11, p. 32). This emphasizes the importance of the question of forces active during replication and genome segregation.

Comparative examination of bacteria and Cyanophyceae and continued research on certain eucaryotic organisms appear highly desirable for the development of a definitive concept of the replication of nuclear structures in protocaryotic organisms.

Bibliography

ABELSON, J., and C. A. THOMAS Jr., 1966: The anomaly of the T 5 bacteriophage DNA molecule. J. molec. Biol. **18**, 262.

ABRAMS, A., L. NIELSEN, and J. THAEMERT, 1964: Rapidly synthesized ribonucleic acid in membrane ghosts from *Streptococcus faecalis* protoplasts. Biochim. biophys. Acta **80**, 325.

ADAMS, J. N., 1963: Nuclear morphogenesis during the developmental cycles of some members of the genus *Nocardia*. J. gen. Microbiol. **33**, 429.

ADLER, H. I., W. D. FISHER, A. COHEN, and A. A. HARDIGREE, 1967: Miniature *Escherichia coli* cells deficient in DNA. Proc. nat. Acad. Sci. U.S.A. **57**, 321.

— and A. A. HARDIGREE, 1965: Growth and division of filamentous forms of *Escherichia coli*. J. Bact. **90**, 223.

AGENO, M., E. DORE, C. FRONTALI, M. ARCA, L. FRONTALI, and G. TECCE, 1965: Interaction between denatured DNA and RNA of *B. stearothermophilus* involving only one half of total DNA. Atti Accad. nazion. Lincei, R. C. Cl. Sci. fis. mat. nat. **38**, 325.

ALFOLDI, L., G. S. STENT, and R. C. CLOWES, 1962: The chromosomal site of the RNA control (RC) locus in *Escherichia coli*. J. molec. Biol. **5**, 348.

AMES, B. N., and D. T. DUBIN, 1960: The role of polyamines in the neutralization of bacteriophage deoxyribonucleic acid. J. biol. Chem. **235**, 769.

— — and S. M. ROSENTHAL, 1958: Presence of polyamines in certain bacterial viruses. Science **127**, 814.

ANDERSON, T. F., et R. MAZÉ, 1957: Analyse de la descendance de zygotes formés par conjugaison chez *Escherichia coli* K 12. Ann. Inst. Pasteur **93**, 194.

— E. L. WOLLMAN et F. JACOB, 1957: Sur les processus de conjugaison et de recombination chez *E. coli* III. Aspects morphologiques en microscopie électronique. Ann. Inst. Pasteur **93**, 450.

ANGELOV, S., N. SPASOVA, I. KUJUMDŽIEV, S. GALABOV, i P. NIKOLOV, 1963: In Bulgarian (Submicroscopic structure and amino acid composition of experimentally obtained large bodies of bacteria). Bolg. Akad. Nauk, Izvest. mikrobiol. Inst. **15**, 79.

ANTOHI, S., I. MORARU et A. PETROVICI 1966: Un système de conjugaison hétérospécifique de haute fréquence entre la souche 31 A de staphylocoque et la souche 207 de *Bacillus megaterium*. C. R. Acad. Sci. (Paris) D **262**, 220.

ARMSTRONG, R. L., and J. A. BOEZI, 1965: Studies of *Escherichia coli* ribonucleic acid — deoxyribonucleic acid complex. Biochim. biophys. Acta **103**, 60.

ARONSON, A. I., and S. SPIEGELMAN, 1961: Protein and ribonucleic acid synthesis in a chloramphenicol-inhibited system. Biochim. biophys. Acta **53**, 70.

ATKINS, H., 1964: Topological considerations of the DNA molecule. Guy's Hosp. Rep. **113**, 74.

AVERY, O. T., C. M. MacLEOD, and M. McCARTY, 1944: Studies on the chemical nature of the substance inducing transformation in pneumococcal cells; Induction of transformation by a desoxyribonucleic acid fraction isolated from Pneumococcus type III. J. exp. Med. **79**, 137.

BACHRACH, U., and I. COHEN, 1961: Spermidine in the bacterial cell. J. gen. Microbiol. **26**, 1.

BADIAN, J., 1930: Z cytologji miksobakteryj. Acta Soc. bot. polon. **7**, 55.

— 1933: Eine cytologische Untersuchung über das Chromatin und den Entwicklungszyklus der Bakterien. Arch. Mikrobiol. **4**, 409.

BAGHAVAN, N. V., and W. ATCHLEY, 1965: Properties of a deoxyribonucleoprotein complex derived from *Bacillus subtilis*. Biochemistry **4**, 234.

BAHR, G. F., 1954: Osmium tetroxide and ruthenium tetroxide and their reactions with biologically important substances. Exp. Cell Res. **7**, 457.

BAHR, H., and W. SCHWARTZ, 1957: Vergleichende cytologische Untersuchungen an farblosen fädigen Schwefelmikroben und an hormogonalen Cyanophyceen. Biol. Zbl. **76**, 183.

BALDWIN, R. L., and E. M. SHOOTER, 1963: On the replication of deoxyribonucleic acid in *Escherichia coli*. Biochem. J. **89**, 9 P.

BAMANN, E., and H. TRAPMANN, 1959: Durch Metall-Ionen katalysierte Vorgänge, vornehmlich im Bereich der seltenen Erdmetalle. Advanc. Enzymol. **21**, 169.

BANERJEE, K. C., and D. J. PERKINS, 1962: The interactions of some metal ions with deoxyribonucleic and ribonucleic acids in aqueous solution. Biochim. biophys. Acta **61**, 1.

BARER, R., 1952: Interference microscopy and mass determination. Nature **169**, 366.
— and S. JOSEPH, 1955: Refractometry of living cells III. Techniques and optical methods. Quart. J. micr. Sci. **96**, 423.
— — 1958: Concentration and mass measurements in microbiology. J. appl. Bact. **21**, 146.
— and K. F. A. Ross, 1952: Refractometry of living cells. J. Physiol. **118**, 38 P.
BARNER, HAZEL D., and S. S. COHEN, 1956: Synchronization of division of a thymine-less mutant of *Escherichia coli*. J. Bact. **72**, 115.
— — 1957: The isolation and properties of amino acid requiring mutants of a thymineless bacterium. J. Bact. **74**, 350.
BARON, S. L., 1963: Transfer of episomes between bacterial genera. Trans. N.Y. Acad. Sci. **27**, 999.
— and S. FALKOW, 1961: Genetic transfer of episomes from *Salmonella typhosa* to *Vibrio comma*. Rec. Genet. Soc. Amer. **30**, 59.
BARR, G. C., and J. A. V. BUTLER, 1963: Biosynthesis of nucleic acids in *Bacillus megaterium* II. The formation of ribonucleic acid by nuclear material *in vitro*. Biochem. J. **88**, 252.
BAUER, L., 1962: Untersuchungen an *Sphaeromyxa xanthochlora* n. sp., einer auf Tropfkörpern vorkommenden Myxobakterienart. Arch. Hyg. **146**, 392.
BAUMANN-GRACE, JOYCE B., and J. TOMCSIK, 1959: Mode of action of trypsin on *Bacillus megaterium*. Proc. Soc. exp. Biol. Med. **101**, 571.
BECK, SIBYLLE, 1963: Licht- und elektronenmikroskopische Untersuchungen an einer sporenbildenden Cyanophycee aus dem Formenkreis von *Pleurocapsa fuliginosa* Hauck. Flora **153**, 194.
BEEBE, J. M., 1941: The morphology and cytology of *Myxococcus xanthus* n. sp. J. Bact. **42**, 193.
BEER, M., and C. R. ZOBEL, 1961: Electron stains II. Electron microscopic studies on the visibility of stained DNA molecules. J. molec. Biol. **3**, 717.
BELOZERSKIJ, A. N., 1939: in Russian (On nuclear substance in bacteria). Mikrobiologija **8**, 504.
— and A. S. SPIRIN, 1960: Chemistry of the nucleic acids of microorganisms, in: The Nucleic Acids, E. CHARGAFF and J. N. DAVIDSON ed., New York: Acad. Press, vol. 3, 147.
BENDICH, A., and H. S. ROSENKRANZ, 1963: Some thoughts on the double-stranded model of deoxyribonucleic acid. Progr. nucleic Acid Res. **1**, 219.
BENNETT, E. O., and R. P. WILLIAMS, 1960: Nucleic acid phosphorus content of *Bacillus anthracis* and *Bacillus cereus*. II. The nucleic acid phosphorus content of the spores. Jap. J. Microbiol. **4**, 157.
BERGERSEN, F. J., 1953 a: A probable growth cycle in *Bacillus megaterium*. J. gen. Microbiol. **9**, 26.
— 1953 b: Cytological changes induced in *Bacterium coli* by chloramphenicol. J. gen. Microbiol. **9**, 353.
BERGH, A. K., S. J. WEBB, and C. S. MCARTHUR, 1965: A histone-like fraction bound to lipid in *Staphylococcus epidermidis*. Canad. J. Biochem. **43**, 623.
BHASKARAN, K., 1958: Genetic recombination in *Vibrio cholerae*. J. gen. Microbiol. **19**, 71.
— 1960: Recombination of characters between mutant stocks of *Vibrio cholerae*, strain 162. J. gen. Microbiol. **23**, 47.
— and S. S. IYER, 1961: Genetic recombination in *Vibrio cholerae*. Nature **189**, 1030.
BILLEN, D., 1959 a: Influence of unbalanced growth on subsequent x-ray induced inhibition of deoxyribonucleic acid synthesis in *Escherichia coli* 15 T—. Nature **184**, 174.
— 1959 b: Alteration in the radiosensitivity of *Escherichia coli* through modification of cellular macromolecular components. Biochim. biophys. Acta **34**, 110.
— 1960 a: Capacity of extracts from prestarved *Escherichia coli* to incorporate thymidine into deoxyribonucleic acid. Nature **187**, 1044.
— 1960 b: Effects of prior alteration in nucleic acid and protein metabolism on subsequent macromolecular synthesis by irradiated bacteria. J. Bact. **80**, 86.
— 1961: Modification of the capacity of bacteria to synthesize deoxyribonucleic acid following x-radiation. An „*in vivo*" and „*in vitro*" study. Sem. Hôp. Pathol. Biol. **9**, 758.
— 1962: Alteration in deoxyribonucleic acid synthesizing capacity in bacteria: an *in vivo-in vitro* study. Biochim. biophys. Acta. **55**, 960.
— 1963: Unbalanced deoxyribonucleic acid synthesis: its role in X-ray induced bacterial death. Biochim. biophys. Acta **72**, 608.

BILLEN, D., 1964: Alteration in the sequence of deoxyribonucleic acid synthesis by thymine deprivation. Exp. Cell Res. 34, 396.
— and R. HEWITT, 1965: Physiological aspects of modification and restoration of chromosomal synthesis in bacteria after x-irradiation. J. Bact. 90, 1218.
— — and G. JORGENSEN, 1965: X-ray induced perturbations in the replication of the bacterial chromosome. Biochim. biophys. Acta 103, 440.
BIRCH-ANDERSEN, A., 1955: Reconstruction of the nuclear sites of Salmonella typhimurium from electron micrographs of serial sections. J. gen. Microbiol. 13, 327.
— O. MAALØE, and F. S. SJÖSTRAND, 1953: High-resolution electron micrographs of sections of E. coli. Biochim. biophys. Acta 12, 395.
BISSET, K. A., 1948 a: Observations upon the bacterial nucleus. J. Hyg. 46, 264.
— 1948 b: The cytology of smooth and rough variation on bacteria. J. gen. Microbiol. 2, 83.
— 1948 c: The cytology of the gram-positive cocci. J. gen. Microbiol. 2, 126.
— 1948 d: Nuclear fusion and reorganization in a Lactobacillus and a Streptococcus. J. gen. Microbiol. 2, 248.
— 1948 e: Nuclear reorganization in non-sporing bacteria. J. Hyg. 46, 173.
— 1949: The nuclear cycle in bacteria. J. Hyg. 47, 182.
— 1950: Observations upon the resting nucleus of the typhoid bacterium. Exp. Cell Res. 1, 473.
— 1952 a: The interpretation of appearances in the cytological staining of bacteria. Exp. Cell Res. 3, 681.
— 1952 b: Spurious mitotic spindles and fusion tubes in bacteria. Nature 169, 347.
— 1952 c: The evidence for mitotic spindles in bacteria. Science 116, 154.
— 1952 d: Complete and reduced life cycles in Rhizobium. J. gen. Microbiol. 7, 233.
— 1952 e: Bacterial cytology. Int. Rev. Cytol. 1, 93.
— 1953 a: The cytology of Caryophanon latum. Mycologia 45, 790.
— 1953 b: Do bacteria have mitotic spindles, fusion-tubes and mitochondria? J. gen. Microbiol. 8, 50.
— 1954 a: The cytology of Micrococcus cryophilus. J. Bact. 67, 41.
— 1954 b: The production of stainable granules in bacteria by the adsorption of DNA against the cell envelopes. Exp. Cell Res. 7, 232.
— JOYCE GRACE, and E. O. MORRIS, 1951: The nuclear reduction process in bacteria. Exp. Cell Res. 2, 388.
— and C. M. F. HALE, 1953: Complex cellular structure in bacteria. Exp. Cell Res. 5, 449.
BISWAS, B. B., 1956: Chemical nature of nucleic acids in Cyanophyceae. Nature 177, 95.
— 1957: Cytochemical studies on the central body of the Cyanophyceae. Cytologia 22, 90.
— 1961: Studies on the nucleoproteins of Nostoc muscorum. Trans. Bose Res. Inst. Calcutta 24, 25.
— and J. MYERS, 1960: Characterization of deoxyribonucleic acid from Anacystis nidulans. Nature 188, 1029.
BLADEN, H. A., Jr., 1963: Demonstration of an unusual ultrastructure found in Bacteroides: a conjugatory bridge? J. Bact. 85, 250.
BLADEN, H. A., R. BYRNE, J. G. LEVIN, and M. W. NIRENBERG, 1965: An electron microscopic study of a DNA-ribosome complex formed in vitro. J. molec. Biol. 11, 78.
BLINKOVA, A. A., S. E. BRESLER and V. A. LANZOV, 1965: DNA synthesis and chromosome transfer in Escherichia coli K 12. Z. Vererbungsl. 96, 267.
BODMER, W. F., 1965: Recombination and integration in Bacillus subtilis transformation: involvement of DNA. J. molec. Biol. 14, 534.
BOIVIN, A., R. TULASNE et R. VENDRELY, 1947 a: Localisation et rôle des deux acides nucléiques chez les bactéries et le problème du noyau bactérien. C. R. Acad. Sci. (Paris) 225, 703.
— — — et R. MINCK, 1947 b: Le noyau des bactéries. Cytologie et cytochimie des bactéries normales et des bactéries traitées par la pénicilline. Arch. Sci. physiol. 1, 307.
— — — — 1948: Le noyau des bactéries. Presse méd. 56, 440.
— — — 1949: Le rôle des deux acides nucléiques dans la constitution et dans la vie de la cellule bactérienne, et plus généralement de toutes les cellules vivantes. Exp. Cell Res. Suppl. 1, 208.
BONHOEFFER, F., and A. GIERER, 1963: On the growth mechanism of the bacterial chromosome. J. molec. Biol. 7, 534.

Boquel. P., et Y. Lehoult, 1948: Action du venin de *Naja tripudians*, de *Vipera aspis* et de *Vipera russellii* (Daboia) sur le cytoplasme des bactéries. Ann. Inst. Pasteur 74, 339.

Bourguignon, Marie-France, 1964: Étude autoradiographique de molécules individuelles d'acide désoxyribonucléique de *Escherichia coli*. Arch. int. Physiol. Biochim. 72, 516.

Bradfield, J. R. G., 1954: Electron microscopic observations on bacterial nuclei. Nature 173, 184.

— 1956: Organization of bacterial cytoplasm, in: Bacterial Anatomy, 6th Sympos. Soc. Gen. Microbiol., Cambridge: Univ. Press, 296.

Bradley, S. G., 1957: Heterokaryosis in *Streptomyces coelicolor*. J. Bact. 73, 581.

— 1959: Mechanisms controlling variation in Streptomycetes. Ann. N.Y. Acad. Sci. 81, 899.

— D. L. Anderson, and L. A. Jones, 1959: Genetic interactions within heterocaryons of Streptomycetes. Ann. N.Y. Acad. Sci. 81, 811.

— and J. Lederberg, 1956 a: Heterocaryosis in *Streptomyces*. Bact. Proc. 48.

— — 1956 b: Heterocaryosis in Streptomyces. J. Bact. 72, 219.

Braendle, D. H., and W. Szybalski, 1957 a: Recombination in balanced heterokaryons of *Streptomyces fradiae*. Bact. Proc., 52.

— — 1957 b: Genetic interaction among Streptomycetes: heterocaryosis and syncaryosis. Proc. nat. Acad. Sci. U.S.A. 43, 947.

— — 1959: Heterocaryotic compatibility, metabolic cooperation, and genetic recombination in *Streptomyces*. Ann. N.Y. Acad. Sci. 81, 824.

Braun, W., 1965: Bacterial Genetics. Philadelphia: Saunders.

Bremer, H., and M. W. Konrad, 1964: A complex of enzymatically synthesized RNA and template DNA. Proc. nat. Acad. Sci. U.S.A. 51, 801.

Brieger, E. M., 1963: Structure and Ultrastructure of Microorganisms. New York: Academic Press.

— and Audrey M. Glauert, 1956: Spore-like structures in the tubercle bacillus. Nature 178, 544.

— and C. F. Robinow, 1947: Demonstration of chromatinic structures in avian tubercle bacilli in the early stages of development. J. Hyg. 45, 413.

Bringmann, G., 1950: Vergleichende licht- und elektronenmikroskopische Untersuchungen an Oszillatorien. Planta 38, 541.

Brinton, C. C. Jr., 1965: The structure, function, synthesis and genetic control of bacterial pili and a molecular model for DNA and RNA transport in gram-negative bacteria. Trans. N.Y. Acad. Sci. 27, 1003.

Britten, R., 1962: Hydrolysis of RNA by lead acetate. C. R. Lab. Carlsberg 32, 271.

Brock, T. D., 1961: Chloramphenicol. Bact. Rev. 25, 32.

Bütschli, O., 1890: Über den Bau der Bakterien und verwandter Organismen. Leipzig.

— 1896: Weitere Ausführungen über den Bau der Cyanophyceen und Bakterien. Leipzig.

— 1900: Notiz über Teilungszustände des Zentralkörpers bei einer Nostocacee. Verh. nat.-med. Ver. Heidelberg N. F. 6, 63.

Burdon, K. L., J. C. Stokes, and C. E. Kimbrough, 1942: Studies of the common aerobic spore-forming bacilli I. Staining for fat with sudan black B—safranin. J. Bact. 43, 717.

Butler, J. A. V., and G. N. Godson, 1963: Biosynthesis of nucleic acids in *Bacillus megaterium* I. The isolation of a nuclear material. Biochem. J. 88, 176.

— — 1964: „Nuclear" and cytoplasmic ribosomes in *B. megaterium*. Nature 201, 876.

Butzow, J. J., and G. L. Eichhorn, 1965: Interactions of metal ions with polynucleotides and related compounds IV. Degradation of polyribonucleotides by zinc and other divalent metal ions. Biopolymers 3, 95.

Byrne, R., J. G. Levin, H. A. Bladen, and M. W. Nirenberg, 1964: The *in vitro* formation of a DNA-ribosome complex. Proc. nat. Acad. Sci. U.S.A. 52, 140.

Cairns, J., 1962 a: Proof that the replication of DNA involves separation of the strands. Nature 194, 1274.

— 1962 b: The application of autoradiography to the study of DNA viruses. Cold Spr. Harb. Symp. quant. Biol. 27, 311.

— 1963 a: The bacterial chromosome and its manner of replication as seen by autoradiography. J. molec. Biol. 6, 208.

— 1963 b: The chromosome of *Escherichia coli*. Cold Spr. Harb. Symp. quant. Biol. 28, 43.

CALDWELL, P. C., and C. HINSHELWOOD, 1950: The nucleic acid content of *Bact. lactis aerogenes.* J. chem. Soc. 1415.
— E. L. HACKOR, and C. HINSHELWOOD, 1950: The ribose nucleic acid content and cell growth of *Bact. lactis aerogenes.* J. chem. Soc. 3151.
CAMPBELL, A. M., 1962: Episomes. Adv. Genet. **11**, 101.
CARO, L. G., 1961: Localization of macromolecules in *Escherichia coli* I. DNA and proteins. J. biophys. biochem. Cytol. **9**, 539.
— and F. FORRO, Jr., 1961: Localization of macromolecules in *Escherichia coli* II. RNA and its site of synthesis. J. biophys. biochem. Cytol. **9**, 555.
— and M. SCHNÖS, 1966: The attachment of the male-specific bacteriophage Fl to sensitive strains of *Escherichia coli.* Proc. nat. Acad. Sci. U.S.A. **56**, 126.
CASSEL, W. A., 1950: The use of perchloric acid in bacterial cytology. J. Bact. **59**, 185.
— 1951 a: A procedure for the simultaneous demonstration of the cell walls and chromatinic bodies of bacteria. J. Bact. **62**, 239.
— 1951 b: Apparent fusion of the chromatinic bodies in species of *Bacillus.* J. Bact. **62**, 514.
— and W. G. HUTCHINSON, 1954: Nuclear studies on the smaller myxophyceae. Exp. Cell Res. **6**, 134.
— — 1955: Fixation and staining of the bacterial nucleus. Stain Technol. **30**, 105.
CATLIN, B. W., 1956: Extracellular deoxyribonucleic acid of bacteria and a deoxyribonuclease inhibitor. Science **124**, 441.
— 1960 a: Interspecific transformation of *Neisseria* by culture slime containing deoxyribonucleate. Science **131**, 608.
— 1960 b: Transformation of *Neisseria meningitidis* by deoxyribonucleates from cells and from culture slime. J. Bact. **79**, 579.
— and L. S. CUNNINGHAM, 1958: Studies of extracellular and intracellular bacterial deoxyribonucleic acids. J. gen. Microbiol. **19**, 522.
CHAN, H. K. H., and K. G. LARK, 1967: Polarity of DNA replication in *Salmonella typhimurium.* Bact. Proc., 57.
CHANCE, H. L., 1951: Staining apparent nuclear material in *Bacillus cereus, Neisseria catarrhalis* and other cocci. Stain Technol. **26**, 77.
— 1952: Crystal violet as a nuclear stain for *Gaffkya tetragena* and other bacteria. Stain Technol. **27**, 253.
— 1954: A nuclear stain for *Escherichia coli* and related organisms. Stain Technol. **29**, 185.
— 1957: Nuclear division in *Escherichia coli* as revealed by acid fuchsin. J. Bact. **74**, 67.
— 1958: Effect of fixatives upon distribution of stainable material in *Escherichia coli* as shown by hydrogen chloride method of staining. J. Bact. **75**, 629.
CHAPMAN, G. B., 1959: Electron microscope observations on the behavior of the bacterial cytoplasmic membrane during cellular division. J. biophys. biochem. Cytol. **6**, 221.
— 1960: Electron microscopy of cellular division in *Sarcina lutea.* J. Bact. **79**, 132.
— and J. HILLIER, 1953: Electron microscopy of ultra-thin sections of bacteria I. Cellular division in *Bacillus cereus.* J. Bact. **66**, 362.
CHARDARD, R., 1960: L'ultrastructure des chromosomes métaphasiques d'une Orchidée. Étude au microscope électronique. C. R. Acad. Sci. (Paris) **250**, 1894.
CHATTERJEE, S. N., and J. K. SARKAR, 1962: Some observations on the vaccinia virus treated with uranyl acetate. Proc. Fifth Internat. Congr. Electron Microscopy, New York: Academic Press 2, S-4.
CHATTON, E., 1918: Les Péridiniens parasites. Morphologie, réproduction, éthologie. Arch. zool. exp. gén. **59**, 1.
— 1937: Titres et Travaux Scientifiques. Sète: Sottano.
— et R. POISSON, 1930: Sur l'existence dans le sang des Crabes de Péridiniens parasites. C. R. Soc. Biol. **105**, 553.
CLARK, D. J., and O. MAALØE, 1967: DNA replication and the division cycle in *Escherichia coli.* J. molec. Biol. **23**, 99.
CLARK, J. B., R. B. GALYEN, and R. B. WEBB, 1953: The effect of organic solvents on the appearance of bacterial nuclei. Stain Technol. **28**, 313.
— and R. B. WEBB, 1955: Ploidy studies on the large cells of *Micrococcus aureus.* J. Bact. **70**, 454.
CLIFTON, C. E., and H. EHRHARDT, 1952: Nuclear changes in living cells of a variant of *Bacillus anthracis.* J. Bact. **63**, 537.

COHEN, A., H. I. ADLER, W. D. FISHER, and F. W. SHULL, 1967: Induction of cell division in *Escherichia coli* filaments. Bact. Proc., 58.

COHEN, S. S., and HAZEL D. BARNER, 1954: Studies on unbalanced growth in *Escherichia coli*. Proc. nat. Acad. Sci. U.S.A. **40**, 885.

— — 1955: Enzymatic adaptation in a thymine requiring strain of *Escherichia coli*. J. Bact. **69**, 59.

COHEN-BAZIRE, GERMAINE, and R. KUNIZAWA, 1963: The fine structure of *Rhodospirillum rubrum*. J. Cell Biol. **16**, 401.

CONTI, S. F., and M. E. GETTNER, 1962: Electron microscopy of cellular division in *Escherichia coli*. J. Bact. **83**, 544.

COOPER, S., and C. HELMSTETTER, 1967: A model of the genome of *Escherichia coli* B/r growing at different rates. Bact. Proc., 57.

COSTERTON, J. W. F., R. G. E. MURRAY, and C. F. ROBINOW, 1961: Observations on the motility and the structure of *Vitreoscilla*. Canad. J. Microbiol. **7**, 329.

COUSSONS, R. T., and R. M. COLE, 1967: Minimal size of colony-forming units of group A streptococcal L forms. Bact. Proc., 70.

CRAVERI, R., L. R. HILL, P. L. MANACHINI, and L. G. SILVESTRI, 1965: Deoxyribonucleic acid base compositions among thermophilic actinomycetes. The occurrence of two strains with low GC content. J. gen. Microbiol. **41**, 335.

CRAWFORD, E. M., and R. F. GESTELAND, 1964: The adsorption of bacteriophage R-17. Virology **22**, 165.

CUMMINGS, D. J., 1965: Macromolecular synthesis during synchronous growth of *Escherichia coli* B/r. Biochim. biophys. Acta **95**, 341.

CUNNINGHAM, L., 1959: Micrococcal nuclease and some products of its action. Ann. N.Y. Acad. Sci. **81**, 788.

DALTON, A. J. 1955: A chrome-osmium fixative for electron microscopy. Anat. Rec. **121**, 281.

DAVIDSON, P. F., D. FREIFELDER, and B. W. HOLLOWAY, 1964: Interruptions in the polynucleotide strands in bacteriophage DNA. J. molec. Biol. **8**, 1.

DE, D. N., and S. N. GHOSH, 1965: Cytochemical evidence for the apparent absence of histone in the cells of Cyanophyceae. J. Histochem. Cytochem. **13**, 298.

DE, N. L., A. GUHA, and N. N. DAS-GUPTA, 1953: Phase contrast and electron microscopic studies on the nuclear apparatus of *Escherichia coli*. Proc. roy. Soc. (London) B **141**, 199.

DEAN, A. C. R., 1962: Nucleic acid and protein content of *Bact. lactis aerogenes*, II. Variation during conditions of nutrient deficiency. Proc. roy. Soc. (London) B **155**, 589.

— and C. HINSHELWOOD, 1960: Variations in the nucleic acid content of *Bact. lactis aerogenes* during the growth cycle. Proc. roy. Soc. (London) B **151**, 348.

DE BOER, WILLEMINA E., J. W. M. LA RIVIÈRE, and A. L. HOUWINK, 1961: Observations on the morphology of *Thiovolum maius* Hinze. Antonie v. Leeuwenhoek **27**, 447.

DEERING, R. A., 1958: Studies on division inhibition and filament formation in *Escherichia coli* by ultraviolet light. J. Bact. **76**, 123.

DE HAAN, P. G., and A. H. STOUTHAMER, 1963: F-prime transfer and multiplication of sex-duced cells. Genet. Res. **4**, 30.

DE HALLER, G., E. KELLENBERGER et C. ROUILLIER, 1964: Étude au microscope électronique des plasmas contenant de l'acide désoxyribonucléique III. Variations ultra-structurales des chromosomes d'*Amphidinium*. J. Microsc. **3**, 627.

DEHORNE, A., 1920: Contribution à l'étude comparée de l'appareil nucléaire des Infusoires, des Euglènes et des Cyanophycées. Arch. Zool. exper. et gén. **60**, 47.

DEKKER, C. A., 1960: Nucleic acids. Selected topics related to their morphology and chemistry. Annu. Rev. Biochem. **29**, 453.

DELAMATER, E. D., 1951: A staining and dehydrating procedure for the handling of microorganisms. Stain Technol. **26**, 199.

— 1953: Aspects of bacteria as cells and as organisms. Int. Rev. Cytol. **2**, 158.

— 1959: A cytological and chemical analysis of the bacterial nucleus I. A method for the isolation of nuclei from *Bacillus megaterium* and the cytology of the isolated structures. Exp. Cell Res. **16**, 636.

— 1962: Withdrawal of the concept of the occurrence of classical mitosis in bacteria. Nature **195**, 309.

— M. E. HUNTER, W. SZYBALSKI, and V. BRYSON, 1955: Chemically induced aberrations of mitosis in bacteria. J. gen. Microbiol. **12**, 203.

— and S. MUDD, 1951: The occurrence of mitosis in the vegetative phase of *Bacillus megaterium*. Exp. Cell Res. **2**, 499.

DELAPORTE, BERTHE, 1934: Sur la structure et le processus de sporulation de l'*Oscillospira guilliermondi*. C. R. Acad. Sci. (Paris) **198**, 1187.
— 1935: Recherches sur la cytologie des bacilles de l'intestin des têtards. C. R. Acad. Sci. (Paris) **201**, 1409.
— 1936 a: Nouvelles recherches sur la cytologie des Bactéries. C. R. Acad. Sci. (Paris) **202**, 1382.
— 1936 b: Recherches cytologiques sur le groupe des Coccacées. C. R. Acad. Sci. (Paris) **203**, 199.
— 1939/1940: Recherches cytologiques, sur les Bactéries et les Cyanophycées. Rev. gén. Bot. **51**, 615, 689, 748; **52**, 40, 75, 112.
— 1940: Observations cytologiques sur *Spirulina versicolor* Cohn. C. R. Acad. Sci. (Paris) **210**, 305.
— 1950: Observations on the cytology of bacteria. Advanc. Genet. **3**, 1.
— 1958: Observations cytologiques sur la structure nucléaire d'une Sarcine de grandes dimensions *Zymosarcina ventriculi*. Rev. Cytol. Biol. vég. **18**, 345.
— 1964 a: Étude comparée de grands spirilles formant des spores: *Sporospirillum* (*Spirillum*) *praeclarum* (Collin) n. gen., *Sporospirillum gyrini* n. sp. et *Sporospirillum bisporum* n. sp. Ann. Inst. Pasteur **107**, 246.
— 1964 b: Étude déscriptive de bactéries de très grandes dimensions. Ann. Inst. Pasteur **107**, 845.
DELBRÜCK, M., and G. S. STENT, 1957: On the mechanism of DNA replication, in: The Chemical Basis of Heredity. W. D. McELROY and B. GLASS eds. Baltimore: Johns Hopkins Press.
DE LEY, J., 1964: Effect of mutation on DNA composition of some bacteria. Antonie v. Leeuwenhoek **30**, 281.
— and J. W. PARK, 1966: Molecular biological taxonomy of some free-living nitrogen-fixing bacteria. Antonie v. Leeuwenhoek **32**, 6.
DE LOZÉ, C., et H. LENORMANT, 1959: Effet des fixateurs sur la nucléohistone de thymus. Bull. Soc. Chim. biol. (Paris) **41**, 337.
DENNIS, E. S., and R. G. WAKE, 1966: Autoradiography of the *Bacillus subtilis* chromosome. J. molec. Biol. **15**, 435.
DÉVIDÉ, Z., 1949: Izvještaj o citolškim istraživanjima leukotiobakterija. Ljetopis jugosl. Akad. Znanosti i Umjetnosti u Zagrebu **55**.
DIENES, L., 1963: Comparative morphology of L forms and PPLO, in: Recent Progress in Microbiology, Symposia held at the Eighth International Congress for Microbiology, Montréal 1962. Toronto: Univ. Toronto Press, 43.
DODGE, J. D., 1960: The nucleus and nuclear divisions in the Dinophyceae. Arch. Protistenk. **106**, 442.
— 1963: Chromosome structure in the Dinophyceae I. The spiral chromonema. Arch. Mikrobiol. **45**, 46.
— 1964 a: Nuclear division of the dinoflagellate *Gonyaulax tamarensis*. J. gen. Microbiol. **36**, 269.
— 1964 b: Chromosome structure in the Dinophyceae II. Cytochemical studies. Arch. Mikrobiol. **48**, 66.
DONACHIE, W. D., 1965: Control of enzyme steps during the bacterial cell cycle. Nature **205**, 1084.
DONNELLAN, J. E., and H. J. HOROWITZ, 1957: Irradiation of dry spores of *Bacillus subtilis* with fast charged particles. Radiation Res. **7**, 71.
DOTY, P., J. MARMUR, J. EIGNER, and C. SCHILDKRAUT, 1960: Strand separation and specific recombination in deoxyribonucleic acids: physical chemical studies. Proc. nat. Acad. Sci. U.S.A. **46**, 461.
DOUDNEY, C. O., 1961 a: Recovery of deoxyribonucleic acid synthesis in bacteria after amino acid starvation. Biochem. biophys. Res. Commun. **5**, 405.
— 1961 b: Recovery of deoxyribonucleic acid synthesis in ultra-violet light-exposed bacteria. Biochem. biophys. Res. Commun. **5**, 410.
— and D. BILLEN, 1961: Incorporation of thymidine into deoxyribonucleic acid by extraction from bacteria exposed to ultra-violet light. Nature **190**, 545.
DOUGHERTY, E. C., 1957: Neologisms needed for structures of primitive organisms. J. Protozool. **4**, Suppl., 14.
DRAKULIĆ, M., and M. ERRERA, 1959: Chloramphenicol-sensitive DNA synthesis in normal and irradiated bacteria. Biochim. biophys. Acta **31**, 459.
DRAWERT, H., 1949/50: Zellmorphologische und zellphysiologische Studien an Cyanophyceen I. Literaturübersicht und Versuche an *Oscillatoria borneti* Zukal. Planta **37**, 161.

Drawert, H., und Ingeborg Metzner-Küster, 1958: Fluoreszenz- und elektronen-mikroksopische Untersuchungen an *Beggiatoa alba* und *Thiothrix* nivea. Mikrobiol. **31**, 422.

Driskell-Zamenhof, P. J., and E. A. Adelberg, 1963: Studies on the chemical nature and size of sex factors of *Escherichia coli* K 12. J. molec. Biol. **6**, 483.

Dubnau, D., I. Smith, P. Morell, and J. Marmur, 1965: Gene conservation in *Bacillus* species I. Conserved genetic and nucleic acid base sequence homologies. Proc. nat. Acad. Sci. U.S.A. **54**, 491.

Dubnau, E., and B. A. D. Stocker, 1964: Genetics of plasmids in *Salmonella typhimurium*. Nature **204**, 1112.

Dugan, P. R., and D. G. Lundgren, 1964: Energy supply for the chemoautotroph *Ferrobacillus ferrooxydans*. J. Bact. **89**, 825.

Duguid, J. P., 1948: The influence of cultural conditions on the morphology of *Bacterium aerogenes* with reference to nuclear bodies and capsule size. J. Path. Bact. **60**, 265.

Dworkin, M., and W. Sadler, 1966: Induction of cellular morphogenesis in *Myxococcus xanthus* I. General description. J. Bact. **91**, 1516.

— and H. Voelz, 1962: The formation and germination of microcysts in *Myxococcus xanthus*. J. gen. Microbiol. **28**, 81.

Ebel, J. P., et J. Colas, 1954: Étude cytochimique des corpuscules métachromatiques des levures et de divers microorganismes. Rapp. et Comm. 8e Congr. Int. Bot. Paris, Sect. 21/27, 36.

Eberle, Helen, and K. G. Lark, 1966: Chromosome segregation in *Bacillus subtilis*. J. molec. Biol. **22**, 183.

— — 1967: Chromosome replication in *Bacillus subtilis* cultures growing at different rates. Proc. nat. Acad. Sci. U.S.A. **57**, 95.

Echlin, P., 1962: Details of the fine structure of the primitive blue-green alga. *Anacystis montana* var. *minor*. Proc. 5th Int. Congr. Electron Microscopy, New York: Academic Press, UU-15.

— 1964: The fine structure of the blue-green alga *Anacystis montana* f. *minor* grown in continuous illumination. Protoplasma **58**, 439.

— and E. D. DeLamater, 1962: A cytological and chemical analysis of the bacterial nucleus II. A study of the cytochemical changes involved during the isolation of bacterial nuclear material by means of bile salts and high-frequency sound waves. Exp. Cell Res. **26**, 229.

— and J. Morris, 1965: The relationship between blue-green algae and bacteria. Biol. Rev. (Cambridge) **40**, 143.

Ecker, R. E., and M. Schaechter, 1963 a: Ribosome content and the rate of growth of *Salmonella typhimurium*. Biochim. biophys. Acta **76**, 275.

— — 1963 b: Bacterial growth under conditions of limited nutrition. Ann. N.Y. Acad. Sci. **102**, 549.

Edelman, M., D. Swinton, J. A. Schiff, H. T. Epstein, and Bernice Zeldin, 1967: Deoxyribonucleic acid of the blue-green algae (Cyanophyta). Bact. Rev. **31**, 315.

Edwards, G. A., and J. Fogh, 1960: Fine structure of pleuropneumonia-like organisms in pure culture and in infected tissue culture cells. J. Bact. **79**, 267.

Edwards, M. R., 1962: Plasmalemma and plasmalemmosomes of *Listeria monocytogenes*. 8th Int. Congr. Microbiol. Montréal, Abst. A 6. 2.

Ehrhart, H., und A. Steigler, 1954: Cytologische Untersuchungen an einem streptomycinabhängigen Stamm von *E. coli*. Z. Hyg. **138**, 561.

Eichhorn, G. L., 1962: Metal ions as stabilizers or destabilizers of the deoxyribonucleic acid structure. Nature **194**, 474.

— and J. J. Butzow, 1965: Interactions of metal ions with polynucleotides and related compounds III. Degradation of polyribonucleotides by lanthanum ions. Biopolymers **3**, 79.

— and P. Clark, 1965: Interactions of metal ions with polynucleotides and related compounds V. The unwinding and rewinding of DNA strands under the influence of copper-II ions Proc. nat. Acad. Sci. **53**, 586.

Eigner, J., H. Boedtker, and G. Michaels, 1961: The thermal degradation of nucleic acids. Biochim. biophys. Acta **51**, 165.

Eigsti, O. J., 1946: Colchicine—a bacterial habitat. Amer. J. Bot. **33**, 218.

Eller, G., and W. H. Beckert, 1965: Nitrogen dioxide fixation in bacterial chromatin studies. J. Bact. **90**, 1710.

Ensign, J. C., and R. S. Wolfe, 1964: Nutritional control of morphogenesis in *Arthrobacter crystallopoietes*. J. Bact. **87**, 924.

ENTZ, G., 1921: Über die mitotische Teilung von *Ceratium hirundinella*. Arch. Protistenk. 43, 416.

EPSTEIN, G. W., E. D. RAVICH-BIRGER, and A. A. SVINKINA, 1936: Contributions to the cytology of the tubercle bacillus. Giorn. Batter. e Immunol. 16, 1.

ERSHOV, F. I., 1963: in Russian (Methods for the cytochemical study of bacterial L forms). Lab. Delo 9, No. 10, 42.

EZEKIEL, D. H., 1961: Increase in ribonucleic acid in the bacterial chromatin body during chloramphenicol treatment. J. Bact. 81, 319.

— 1964: Accumulation of ribonucleic acid in bacterial nuclear preparations during treatment of whole cells with 8-azaguanine, tetracyclines, and other inhibitors. J. Bact. 87, 755.

FALKOW, S., R. V. CITARELLA, J. A. WOHLHIETER, and T. Watanabe, 1966: The molecular nature of the R-factor. J. molec. Biol. 17, 102.

— J. MARMUR, W. F. CAREY, W. M. SPILMAN, and L. S. BARON, 1961: Episomic transfer between *Salmonella typhosa* and *Serratia marcescens*. Genetics 46, 703.

— J. A. WOHLHIETER, R. V. CITARELLA, and L. S. BARON, 1964: Transfer of episomic elements to Proteus I. Transfer of F-linked chromosomal determinants. J. Bact. 87, 209.

FAURET-FRÉMIET, E., et C. ROUILLIER, 1958: Étude au microscope électronique d'une bactérie sulfureuse, *Thiovolum maius* Hinze. Exp. Cell Res. 14, 29.

FELSENFELD, G., and S. HUANG, 1959: The interaction of polynucleotides with cations. Biochim. biophys. Acta 34, 234.

— — 1960: The interaction of polynucleotides with metal ions, amino acids and polyamines. Biochim. biophys. Acta 37, 425.

— — 1961: Some effects of charge and structure upon ionic interactions of nucleic acids. Biochim. biophys. Acta 51, 19.

FEULGEN, R., 1918: Darstellung und Eigenschaften der Thyminsäure. Hoppe-Seylers Z. physiol. Chem. 101, 296.

— 1939: Von der Nuclealfärbung zum Plasmalogen. Schriften der Ludwigs-Universität zu Gießen, Heft 1.

— und H. ROSSENBECK, 1924: Anwendung der Nuclealreaktion auf histologische Präparate. Hoppe-Seylers Z. physiol. Chem. 135, 203.

FIKHMAN, B. A., 1963: in Russian (Immersion medium for phase-contrast microscopy of living bacterial cells). Lab. Delo 9, No. 5, 53.

FITZ-JAMES, P. C., 1953: The structure of spores as revealed by mechanical disruption. J. Bact. 66, 312.

— 1954: The duplication of bacterial chromatin. Interpretations of some cytological and chemical studies of the germinating spores of *Bac. cereus* and *Bac. megaterium*. J. Bact. 68, 464.

— 1955 a: The phosphorus fractions of *Bacillus cereus* and *Bacillus megaterium*. I. A comparison of spores and vegetative cells. Canad. J. Microbiol. 1, 502.

— 1955 b: The phosphorus fractions of *Bacillus cereus* and *Bacillus megaterium*. II. A correlation of the chemical with the cytological changes occurring during spore germination. Canad. J. Microbiol. 1, 525.

— 1958: Studies on the morphology and nucleic acid content of protoplasts of *Bacillus megaterium*. J. Bact. 75, 369.

— 1960: Participation of the cytoplasmic membrane in the growth and spore formation of bacilli. J. biophys. biochem. Cytol. 8, 507.

— 1964 a: Sporulation in protoplasts and its dependence on prior forespore development. J. Bact. 87, 667.

— 1964 b: Electron microscopy of *Bacillus megaterium* undergoing isolation of its nuclear bodies. J. Bact. 87, 1202.

— 1964 c: Fate of the mesosomes of *Bacillus megaterium* during protoplasting. J. Bact. 87, 1483.

— 1965: A consideration of the bacterial membrane as the agent of differentiation. 15th Sympos. Soc. gen. Microbiol., Cambridge: University Press, 369.

— C. F. ROBINOW, and G. H. BERGOLD, 1954: Acid hydrolysis of the spores of *B. cereus;* a correlation of chemical and cytological findings. Biochim. biophys. Acta 14, 346.

— and I. ELIZABETH YOUNG, 1959: Comparison of species and varieties of the genus *Bacillus*. Structure and nucleic acid content of spores. J. Bact. 78, 743.

FLEGEL, H., 1953: Untersuchungen an fluorochromierten Bakterienkulturen. Zbl. Bakt., I. Abt. Orig. 159, 342.

Flewett, T. H., 1948: Nuclear changes in *Bacillus anthracis* and their relation to variants. J. gen. Microbiol. **2**, 325.
Floethmann, E., 1954: Cytologische Untersuchungen an *Azotobacter chroococcum*. Arch. Mikrobiol. **20**, 243.
Forro, F., Jr., 1965: Autoradiographic study of bacterial chromosome replication in amino-acid deficient *Escherichia coli* 15 T—. Biophys. J. **5**, 629.
— and S. A. Wertheimer, 1960: The organization and replication of deoxyribonucleic acid in thymine-deficient strains of *Escherichia coli*. Biochim. biophys. Acta **40**, 9.
Fox, M. S., 1960: Fate of transforming deoxyribonucleate following fixation by transformable bacteria. Nature **187**, 1004.
Franklin, R. M., and N. Granboulan, 1965: High resolution autoradiography of bacteria labelled with tritiated thymidine. J. molec. Biol. **14**, 623.
Fraser, D., and H. R. Mahler, 1962: Studies in partially resolved bacteriophage-host systems. VII. Diamines, dyes, empty phage heads and protoplast-infecting agent. Biochim. biophys. Acta **53**, 199.
Freifelder, D., 1966: Replication of DNA during F"lac transfer. Biochem. biophys. Res. Commun. **23**, 576.
— and P. F. Davison, 1962: Hyperchromocity and strand separation in bacterial DNA. Biophys. J. **2**, 249.
— and A. K. Kleinschmidt, 1965: Single-strand breaks in duplex DNA of coliphage T 7 as demonstrated by electron microscopy. J. molec. Biol. **14**, 271.
Friesen, J. D., and O. Maaløe, 1965: On the control of deoxyribonucleic acid synthesis by amino acids in *Escherichia coli*. Biochim. biophys. Acta **95**, 436.
Fruchter, J., 1960: Cu puivire la nucleul bacteriei celulozolitice diu *Sporocytophaga*. An. Univ. C. I. Parhon, Ser. Sti. nat. **9**, No. 24, 61.
Fuerst, C. R., F. Jacob, et E. L. Wollman, 1956: Déterminations de liaisons génétiques, chez *Escherichia coli* K 12, à l'aide de radiophosphore. C. R. Acad. Sci. (Paris) **243**, 2162.
— and G. S. Stent, 1956: Inactivation of bacteria by decay of incorporated radioactive phosphorus. J. gen. Physiol. **40**, 73.
Fuhs, G. W., 1958: Bau, Verhalten und Bedeutung der kernäquivalenten Strukturen bei *Oscillatoria amoena* (Kütz.) Gomont. Arch. Mikrobiol. **28**, 270.
— 1963 a: Addendeum to: E. Kellenberger: Organization of the genetic material of phage, bacteria, and dinoflagellate. Eleventh Int. Congr. Genetics, Proceedings. Oxford: Pergamon Press, **2**, 319.
— 1963 b: Cytochemisch-elektronenmikroskopische Lokalisierung der Ribonukleinsäure und des Assimilats in Cyanophyceen. Protoplasma **56**, 178.
— 1964: Die Wirkung von Uranylsalzen auf die Struktur des Bakteriennucleoids, 1. Mitteilung über Nucleoide. Arch. Mikrobiol. **49**, 383.
— 1965 a: Der physikalische Status der Deoxyribonucleinsäure im Bakterinnucleoid, 2. Mitteilung über Nucleoide. Antonie v. Leeuwenhoek **31**, 25.
— 1965 b: Grundzüge der Nucleoidfeinstruktur, 3. Mitteilung über Nucleoide. Arch. Mikrobiol. **50**, 25.
— 1965 c: Zum chemischen Status der Deoxyribonucleinsäure im Bakteriennucleoid, 4. Mitteilung über Nucleoide. Arch. Mikrobiol. **52**, 91.
— 1965 d: Fine structure and replication of bacterial nucleoids (Symposium on the fine structure and replication of bacteria and their parts, 64th Annual Meeting, American Society for Microbiology). Bact. Rev. **29**, 277.
— 1967: Cytology of blue-green algae: Light microscopic aspects. Int. Sympos. "Algae, Man and the Environment" (D. F. Jackson, ed.). Syracuse: Syracuse Univ. Press (to appear in 1968).
Fukushi, K., 1959: Electron microscopic studies of tubercle bacilli V. Studies on fixation and ultrathin sectioning. Sci. Rep. Res. Inst. Tohoku Univ. Ser. C., **9**, 1.

Gallant, J. A., 1962: Sublethal thymineless damage in *Escherichia coli* B 3. Biochim. biophys. Acta **61**, 302.
— and S. Suskind, 1962: Ribonucleic acid and thymineless death. Biochim. biophys. Acta **55**, 627.
Ganesan, A. T., 1963: Physical and biological studies of *Bacillus subtilis* deoxyribonucleic acid. Ph.D. Thesis, Stanford Univ.
— and J. Lederberg, 1956 a: Membrane-bound fraction of bacterial DNA. Biophys. Soc. Ninth Annu. Meetg., p. 162.
— — 1965 b: A cell-membrane bound fraction of bacterial DNA. Biochim. biophys. Res. Commun. **18**, 824.

GARDNER, N. L., 1906: Cytological studies in Cyanophyceae. Univ. Calif. Publ. Bot. **2**, 237.

GAUSE, G. G., N. P. LOSHKAREVA, I. B. ZBARSKY, and G. F. GAUSE, 1964: Deoxyribonucleic acid base composition in certain bacteria and their mutants with impaired respiration. Nature **203**, 598.

GEITLER, L., 1923: Der Zellbau von *Glaucocystis nostochinearum* and *Gloeochaete mittrockiana* und die Chromatophoren-Symbiosetheorie von MERESCHKOWSKY. Arch. Protistenk. **47**, 1.

— 1936: Schizophyceen, in: LINSBAUERS Handbuch der Pflanzenanatomie, vol. VI/I, B, Berlin: Borntraeger.

— 1942: Schizophyta, in: Natürliche Pflanzenfamilien. Leipzig: Engelmann. — Nachdruck 1959: Berlin: Duncker & Humblot.

— 1953: Endomitose und endomitotische Polyploidisierung. Protoplasmatologia, Handbuch der Protoplasmaforschung, VI/C. Wien: Springer.

— 1959 a: Syncyanosen, in: RUHLANDS Handbuch der Pflanzenphysiologie, **11**, 530, Berlin-Göttingen-Heidelberg: Springer.

— 1959 b: Eine neue Endocyanose, *Cyanoptyche dispersa* n. sp. und Bemerkungen über ähnliche Syncyanosen. Österr. bot. Z. **106**, 464.

— 1960: Schizophyceen, in: Encyclopedia of Plant Anatomy/Handbuch der Pflanzenanatomie, K. LINSBAUER ed., 2nd ed., vol. VI, pt. I., Berlin: Borntraeger.

GIESBRECHT, P., 1957: Zytologische Untersuchungen an Bakterien. Thesis, Math.-Naturw. Fak. Univ. Bonn.

— 1958: Zur Struktur des Bakterienzellkerns. Naturwissenschaften **45**, 473.

— 1959: Elektronenmikroskopische Darstellung verschiedener chemischer und physikalisch-chemischer Reaktionen an den DNS-Strukturen eines Chromosoms (Die Bakterienzelle als „biologisches Reagenzglas"). Zbl. Bakt., I. Abt. Orig. **176**, 413.

— 1961: Über das „Supercoiling"-System der Chromosomen von Bakterien und Flagellaten und seine Beziehungen zu Nucleolus und Kerngrundsubstanz. Zbl. Bakt., I. Abt. Orig. **183**, 1.

— 1962: Vergleichende Untersuchungen an den Chromosomen des Dinoflagellaten *Amphidinium elegans* und denen der Bakterien. Zbl. Bakt., I. Abt. Orig. **187**, 452.

— 1965: Über das Ordnungsprinzip in den Chromosomen von Dinoflagellaten und Bakterien. Zbl. Bakt., I. Abt. Orig. **196**, 516.

— und G. PIEKARSKI, 1958: Zur Organisation des Zellkernes von *Bacillus megaterium*. Arch. Mikrobiol. **31**, 68.

GILLIES, N. E., and T. ALPER, 1960: The nucleic acid content of *Escherichia coli* strains B and B/r. Biochim. biophys. Acta **43**, 182.

GINSBERG, D. M., and J. JAGGER, 1965 a: Evidence that initial ultraviolet damage in *Escherichia coli* strain 15 T—A—U— is independent of growth phase. J. gen. Microbiol. **40**, 171.

— — 1965 b: Radiation sensitivity and growth characteristics of an arginine- and uracil-starved culture of *Escherichia coli* strain 15 T—A—U—. J. gen. Microbiol. **40**, 185.

GLAUERT, AUDREY M., 1962: The fine structure of bacteria. Brit. med. Bull. **18**, 295.

— and E. M. BRIEGER, 1956: Ultra-thin sections of avian tubercle bacilli in a new embedding medium. Proc. Conf. Electron Microscopy (Stockholm), 111.

— and JENNIFER M. ALLEN, 1961: The fine structure of vegetative cells of *Bacillus subtilis*. Exp. Cell Res. **22**, 73.

— and D. A. HOPWOOD, 1961: The fine structure of *Streptomyces violaceoruber* (*S. coelicolor*) III. The walls of the mycelium and spores. J. biophys. biochem. Cytol. **10**, 505.

GODSON, G. N., and J. A. V. BUTLER, 1962: Preparation of a nuclear fraction from *Bacillus megatherium* and its role in the biosynthesis of ribonucleic acid. Nature **193**, 655.

— — 1963: Biosynthesis of nucleic acids in *Bacillus megaterium* III. Biosynthesis of ribonucleic acid *in vivo*. Biochem. J. **88**, 259.

— — 1964: Biosynthesis of nucleic acids in *Bacillus megaterium* IV. Roles of the "nuclear" cytoplasmic and cytoplasmic-membrane components of the cell in the biosynthesis of ribonucleic acid. Biochem. J. **93**, 573.

GOGOL, E. M., and E. ROSENBERG, 1964: Observations on the regulation of DNA biosynthesis. Biochem. biophys. Res. Commun. **14**, 565.

GOLDSCHMIDT, R., 1905: Die Chromidien der Protozoen. Arch. Protistenk. **5**, 126.

GOLDSTEIN, A., and B. J. BROWN, 1961: Effect of sonic oscillation upon "old" and "new" nucleic acids in *Escherichia coli*. Biochim. biophys. Acta **53**, 19.

GOLDSTEIN, A., DORA B. GOLDSTEIN, B. J. BROWN, and SHAO-CHIA CHOU, 1959: Amino-acid starvation in an *Escherichia coli* auxotroph. I. Effects on protein and nucleic acid synthesis and on cell division. Biochim. biophys. Acta **36**, 163.
— J. B. KIRSCHBAUM, and A. ROMAN, 1965: Direction of synthesis of messenger RNA in cells of *Escherichia coli*. Proc. nat. Acad. Sci. **54**, 1669.
GOODGAL, S. H., and N. E. MELECHEN, 1960: Synthesis of transforming DNA in the presence of chloramphenicol. Biochem. biophys. Res. Commun. **3**, 114.
GRACE, JOYCE B., 1951: The life cycle of *Sporocytophaga*. J. gen. Microbiol. **5**, 519.
GRELL, K. G., 1953: Die Chromosomen von *Aulacantha scolymantha* Haeckel. Arch. Protistenk. **99**, 1.
— 1964: The protozoan nucleus, in: The Cell, J. BRACHET and A. E. MIRSKY ed., New York: Academic Press, **6**, 1.
— und G. SCHWALBACH, 1965: Elektronenmikroskopische Untersuchungen an den Chromosomen der Dinoflagellaten. Chromosoma **17**, 230.
— und K. E. WOHLFARTH-BOTTERMANN, 1957: Licht- und elektronenmikroskopische Untersuchungen an dem Dinoflagellaten *Amphidinium elegans* n. sp. Z. Zell-forsch. **47**, 7.
GROS, F., et FRANÇOISE GROS, 1958: Rôle des acides aminés dans la synthèse des acides nucléiques chez *Escherichia coli*. Exp. Cell Res. **14**, 104.
GROSS, J. D., and L. G. CARO, 1965: Genetic transfer in bacterial mating. What mechanism insures the orderly transfer of DNA from donor to recipient cells. Science **150**, 1679.
— — 1966: DNA transfer in bacterial conjugation. J. molec. Biol. **16**, 269.
GRULA, E. A., and M. M. GRULA, 1962: Cell division in a species of *Erwinia*. III. Reversal of inhibition of cell division caused by D-amino acids, penicillin, and ultraviolet light. IV. Metabolic blocks in pantothenate biosynthesis and their relationship to inhibition of cell division. J. Bact. **83**, 981, 989.
— and G. L. SMITH, 1965: Cell division in a species of *Erwinia*. IX. Electron microscopy of normally dividing cells. J. Bact. **90**, 1054.
GUEST, J. R., and C. YANOFSKY, 1966: Relative orientation of gene, messenger and polypeptide chain. Nature **210**, 799.
GUHA, A., N. N. DASGUPTA, and M. L. DE, 1954: Nuclear apparatus of *E. coli* after fixation in chromic acid and osmium tetroxide. J. Bact. **67**, 292.
GUILLIERMOND, A., 1906: Contribution à l'étude cytologique des Cyanophycées. Rev. gén. Bot. **18**, 392, 447.
— 1925: Nouvelles observations sur la structure des Cyanophycées. C. R. Acad. Sci. (Paris) **180**, 951.
— 1926: Nouvelles recherches sur la structure des Cyanophycées. Rev. gén. Bot. **38**, 129, 177.
— 1933: La structure des Cyanophycées. C. R. Acad. Sci. Paris. **197**, 182.

HAGEDORN, H., 1955: Beiträge zur Cytologie und Morphologie der Actinomyceten. Zbl. Bakt., II. Abt. **108**, 353.
— 1959 a: Licht- und elektronenmikroskopische Untersuchungen an *Nocardia coral-lina* (BERGEY et al. 1923). Zbl. Bakt., II. Abt. **112**, 214.
— 1959 b: Elektronenmikroskopische Untersuchungen über den Teilungsverlauf bei *Nocardia corallina* (BERGEY et al. 1923). Zbl. Bakt., II. Abt. **112**, 359.
— 1960: Elektronenmikroskopische Untersuchungen an Blaualgen. Naturwissen-schaften **47**, 430.
— 1961: Untersuchungen über die Feinstruktur der Blaualgenzellen. Z. Naturf. B **16**, 825.
HAHN, F. E., M. SCHAECHTER, W. S. CEGLOWSKI, H. E. HOPPS, and J. CIAK, 1957: Inter-relationships between nucleic acid and protein biosynthesis. I. Synthesis and fate of bacterial nucleic acids during exposure to and recovery from the action of chloramphenicol. Biochim. biophys. Acta **26**, 469.
— and A. D. WOLFE, 1962: Mode of action of chloramphenicol. VIII. Resemblance between labile chloramphenicol RNA and DNA of *Bacillus cereus*. Biochem. biophys. Res. Commun. **6**, 464.
HALE, C. M. F., 1954: Short note on the production of nuclear artifacts resembling mitotic figures in microorganisms by treatment with organic solvents. Exp. Cell Res. **6**, 243.
— 1958: The simultaneous demonstration of cellular structure and flagella in bac-teria. Lab. Practice **7**, 361.
HALL, B. D., M. GREEN, A. P. NYGAARD, and J. BOEZI, 1963: The copying of DNA in T2-infected *E. coli*. Cold Spring Harbor Sympos. quant. Biol. **28**, 201.

HALL, W. T., and G. CLAUS, 1963: Ultrastructural studies on the blue-green algal symbiont in *Cyanophora paradoxa* Kor. J. Cell Biol. **19**, 551.

HALVORSON, H. O., 1967: Developmental stages during outgrowth of bacterial spores. Sympos. Molec. and Cellul. Aspects of Differentiation and Morphogenesis I, Amer. Soc. Microbiol. 67th Annu. Meetg.

HAMMARSTEN, E., 1924: Zur Kenntnis der biologischen Bedeutung der Nucleinsäureverbindungen. Biochem. Z. **144**, 383.

HANAWALT, P. C., 1963: Involvement of synthesis of RNA in thymineless death. Nature **198**, 286.

— 1966: The UV sensitivity of bacteria: Its relation to the DNA replication cycle. Photochem. and Photobiol. **5**, 1.

— O. MAALØE, D. J. CUMMINGS, and M. SCHAECHTER, 1961: The normal DNA replication cycle II. J. molec. Biol. **3**, 156.

— and D. S. RAY, 1964: Isolation of the growing point in the bacterial chromosome. Proc. nat. Acad. Sci. U.S.A. **52**, 125.

— and R. WAX, 1964: Transcription of a repressed gene: Evidence that it requires DNA replication. Science **145**, 1061.

HAROLD, F. M., and Z. Z. ZIPORIN, 1958: The relationship between the synthesis of DNA and protein in *Escherichia coli* treated with sulfur mustard. Biochim. biophys. Acta **28**, 492.

— — 1959: Synthesis of protein and DNA in *Escherichia coli* irradiated with ultraviolet light. Biochim. biophys. Acta **29**, 439.

HARTMAN, P. E., and J. J. PAYNE, 1954: Direct staining of the two types of nucleoproteins in *E. coli*. J. Bact. **68**, 237.

HARTMANN, MAX, 1911: Die Konstitution der Protistenkerne und ihre Bedeutung für die Zellenlehre. Jena: Fischer.

— 1953: Allgemeine Biologie. 4. ed. Stuttgart: Fischer.

HASHIMOTO, T., and P. GERHARDT, 1960: Monochromatic ultraviolet microscopy of microorganisms: Preliminary observations on bacterial spores. J. biophys. biochem. Cytol. **7**, 195.

— and H. B. NAYLOR, 1958: Studies of the fine structure of microorganisms. II. Electron microscopic studies on sporulation of *Clostridium sporogenes*. J. Bact. **75**, 647.

HAUPT, A. W., 1923: Cell structure and cell division in the Cyanophyceae. Bot. Gaz. **75**, 170.

HAYASHI, M., 1965: A DNA-RNA complex as an intermediate of *in vitro* genetic transcription. Proc. nat. Acad. Sci. U.S.A. **54**, 1736.

— M. N. HAYASHI, and S. SPIEGELMAN, 1963 a: Restriction of *in vivo* genetic transcription to one of the complementary strands of DNA. Proc. nat. Acad. Sci. U.S.A. **50**, 664.

— — — 1964: DNA circularity and the mechanism of strand selection in the generation of genetic messages. Proc. nat. Acad. Sci. U.S.A. **51**, 351.

— S. SPIEGELMAN, N. C. FRANKLIN, and S. E. LURIA, 1963 b: Separation of the RNA message transcribed in response to a specific inducer. Proc. nat. Acad. Sci. U.S.A. **49**, 729.

HAYASHI, M. N., and M. HAYASHI, 1966: Participation of a DNA-RNA hybrid complex in *in vivo* genetic transcription. Proc. nat. Acad. Sci. U.S.A. **55**, 635.

HAYES, W., 1953: Observations on a transmissible agent determining sexual differentiation in *Bacterium coli*. J. gen. Microbiol. **8**, 72.

— 1964: The genetics of bacteria and their viruses, studies in basic genetics and molecular biology. Oxford: Blackwell.

HEDÉN, C.-G., 1951: Studies of the infection of *E. coli* with the bacteriophage T 2. Acta path. microbiol. scand. Suppl. **88**.

HEGLER, R., 1901: Untersuchungen über die Organisation der Phycochromaceenzellen. Jahrb. wiss. Bot. **36**, 229.

HELMSTETTER, C. E., 1957: Rate of DNA synthesis during the division cycle of *Escherichia coli* B/r. J. molec. Biol. **24**, 417.

— and S. COOPER, 1967: Rate of DNA synthesis during the division cycle of rapidly growing *Escherichia coli* B/r. Bact. Proc., 57.

— and D. J. CUMMINGS, 1964: An improved method for the selection of bacterial cells at division. Biochim. biophys. Acta **82**, 608.

HERBST, E. J., R. H. WEAVER, and D. C. KEISTER, 1958: The Gram reaction and cell composition: diamines and polyamines. Arch. Biochem. Biophys. **75**, 171.

HERBST, F., 1952: Strahlenbiologische und cytologische Untersuchungen an Blaualgen. Diss. phil. Fak. Univ. Köln, 1952.

HERMAN, R. K., and F. FORRO, Jr., 1964: Autoradiographic study of transfer of DNA during bacterial conjugation. Biophys. J. **4**, 335.

HERSHEY, A. D., and E. BURGI, 1965: Complementary structure of interacting sites and the ends of lambda DNA molecules. Proc. nat. Acad. Sci. U.S.A. **53**, 325.

— — and L. INGRAHAM, 1963: Cohesion of DNA molecules isolated from phage lambda. Proc. nat. Acad. Sci. U.S.A. **49**, 748.

— A. GAREN, D. K. FRASER, and J. D. HUDIS, 1954: Growth and inheritance in bacteriophage. Carnegie Inst. of Washington, Yearbook **53**, 210.

— and N. E. MELECHEN, 1957: Synthesis of phage-precursor nucleic acid in the presence of chloramphenicol. Virology **3**, 207.

HERTWIG, R., 1902: Die Protozoen und die Zelltheorie. Arch. Protistenk. **1**, 1.

HEUMANN, W., 1956: Der Sexualzyklus sternbildender Bakterien. Arch. Mikrobiol. **24**, 362.

— 1960 a: Versuche zur Rekombination sternbildender Bakterien. Naturwiss. **47**, 330.

— 1960 b: Kreuzungsversuche mit sternbildenden Bakterien. Arch. Mikrobiol. **36**, 244.

— 1962 a: Das genetische Verhalten sternbildender Bakterien. Naturwiss. **49**, 430.

— 1962 b: Genetische Untersuchungen sternbildender Bakterien. Z. Vererbungsl. **93**, 441.

— 1963: Genetische Untersuchungen von Bodenbakterien. Zbl. Bakt., I. Abt. Orig. **191**, 426.

— und RUTH MARX, 1964: Feinstruktur und Funktion der Fimbrien bei dem stern bildenden Bakterium *Pseudomonas echinoides*. Arch. Mikrobiol. **47**, 325.

HEWITT, R., and D. BILLEN, 1965: Reorientation of chromosome replication after exposure to ultraviolet light of *Escherichia coli*. J. mol. Biol. **13**, 40.

HIATT, S., 1965: Effects of cupric ions on thermal denaturation of nucleic acids. J. molec. Biol. **11**, 672.

HICKMAN, D. D., and A. W. FRENKEL, 1959: The structure of *Rhodospirillum rubrum*. J. biophys. biochem. Cytol. **6**, 277.

HIERONYMUS, G., 1892: Beiträge zur Morphologie und Biologie der Algen. Cohn's Beitr. Biol. Pflanzen **5**, 461.

HIGASHI, N., 1959: Electron microscopic studies on the organization in the bacterial cell and on the intracellular rickettsial cell. Annu. Rept. Inst. Virus Res. Kyoto Univ. Ser. B **2**, 167.

HILL, L. R., 1966: An index to deoxyribonucleic acid base compositions of bacterial species. J. gen. Microbiol. **44**, 419.

HILLIER, J., S. MUDD, and A. G. SMITH, 1949: Internal structure and nuclei in cells of *Escherichia coli* as shown by improved electron microscope techniques. J. Bact. **57**, 319.

HIRANO, T., 1959: in Japanese (Cytological studies of bacteria III. Nucleus and autogamy in *Bacillus megaterium*). Jap. J. Genet. **34**, 88.

— 1961: Observation of nuclear fusion in living cells of *Bacillus megaterium*. Antonie v. Leeuwenhoek **27**, 457.

HIROTA, Y., 1960: The effect of acridine dyes on mating type factors in *E. coli*. Proc. nat. Acad. Sci. U.S.A. **46**, 57.

HIRSCH, P., and S. F. CONTI, 1964: Biology of budding bacteria I. Enrichment, isolation and morphology of *Hyphomicrobium* spp. Arch. Mikrobiol. **48**, 339.

HOENIGER, JUDITH F. M., 1966: Cellular changes accompanying the swarming of *Proteus mirabilis* II. Observations of stained organisms. Canad. J. Microbiol. **12**, 113.

— and P. F. STUART, 1967: Cytology of spore formation in *Clostridium perfringens* type D. Bact. Proc., 28.

HOFFMAN, H., 1951: Cytochemistry of bacterial nuclear structures. J. Bact. **62**, 561.

HOFFMANN-BERLING, H., D. A. MARVIN und H. DÜRWALD, 1963: Ein fädiger DNS-Phage (fd) und ein sphärischer RNS-Phage (fr) wirtsspezifisch für männliche Stämme von *E. coli*. I. Präparation und chemische Eigenschaften von fd und fr. Z. Naturf. B **18**, 876.

— and R. MAZÉ, 1964: Release of male-specific bacteriophages from surviving host bacteria. Virology **22**, 305.

HOLLANDE, Mme G., 1962: Essai d'interprétation de la structure des bactéries d'après les diverses conceptions des auteurs. Arch. Protistenk. **106**, 101.

HOLLOM, S., and R. H. PRITCHARD, 1965: Effect of inhibition of DNA synthesis in mating on *E. coli* K 12. Genet. Res. **6**, 479.

HOLLOWAY, B. W., 1955: Genetic recombination in *Pseudomonas aeruginosa*. J. gen. Microbiol. **13**, 572.

— 1956: Self-fertility in *Pseudomonas aeruginosa*. J. gen. Microbiol. **15**, 221.

HOLLOWAY, B. W., and B. FARGIE, 1960: Fertility factor and genetic linkage in *Pseudomonas aeruginosa*. J. Bact. **80**, 362.
— and P. A. JENNINGS, 1958: An infectious fertility factor for *Pseudomonas aeruginosa*. Nature **181**, 855.
HOPWOOD, D. A., 1959: Linkage and the mechanism of recombination in *Streptomyces coelicolor*. Ann. N.Y. Acad. Sci. **81**, 887.
— 1960: Phase-contrast observations on *Streptomyces coelicolor*. J. gen. Microbiol. **22**, 295.
— 1967: Genetic analysis and genome structure in *Streptomyces coelicolor*. Bact. Rev. **31**, 373.
— and AUDREY M. GLAUERT, 1960 a: Observations on the chromatinic bodies of *Streptomyces coelicolor*. J. biophys. biochem. Cytol. **8**, 257.
— — 1960 b: The fine structure of *Streptomyces coelicolor*. II. The nuclear material. J. biophys. biochem. Cytol. **8**, 267.
— — 1960 c: The fine structure of the nuclear material of a blue-green alga. *Anabaena cylindrica* Lemm. J. biophys. biochem. Cytol. **8**, 813.
HORVATH, Y., 1963: Heterocaryosis and recombination in *Streptomycetes*. Excerpta med., Int. Congr. Ser., No. **59**, 78.
HUANG, S. L., and G. FELSENFELD, 1960: Solubility of complexes of polynucleotides with spermine. Nature **188**, 301.
HUEBSCHMAN, C., 1952: A method for varying the average number of nuclei in the conidia of *Neurospora crassa*. Mycologia **44**, 599.
HUNTER, M. E., and E. D. DE LAMATER, 1955: Further observations on the bacterial spore nucleus. J. Bact. **69**, 108.
HURST, A., and D. TAYLOR, 1965: Growth inhibition of *Escherichia coli* by some basic proteins prepared from the same strain. Nature **207**, 438.
HUXLEY, H. E., and G. ZUBAY, 1961: Preferential staining of nucleic acid-containing structures for electron microscopy. J. biophys. biochem. Cytol. **11**, 273.

IMSHENETSKY, A. A., and V. V. ALFEROV, 1962: Electron microscopic study of the nuclei in a *Sorangium species*. J. gen. Microbiol. **27**, 391.
ISHIBASHI, M., 1966: Effect of thymine starvation on chromosome transfer during conjugation in *Escherichia coli*. Jap. J. Genet. **41**, 75.

JACKSON, B., and F. I. DESSAU, 1955: Streptococcal desoxyribonuclease for the removal of Feulgan-stainable material. Stain Technol. **30**, 9.
JACOB, F., et S. BRENNER, 1963: Sur la régulation de la synthèse du DNA chez les bactéries: l'hypothèse du réplicon. C. R. Acad. Sci. (Paris) **256**, 298.
— — and F. CUZIN, 1963: On the regulation of DNA replication in bacteria. Cold Spr. Harb. Symp. quant. Biol. **28**, 329.
— and J. MONOD, 1961: On the regulation of the gene activity. Cold Spr. Harb. Symp. quant. Biol. **26**, 193.
— — 1963: Elements of regulatory circuits in bacteria. UNESCO Sympos. biol. Org., New York: Academic Press.
— ANTOINETTE RYTER, and F. CUZIN, 1966: On the association between DNA and membrane in bacteria. Proc. Roy. Soc. (London) **B 164**, 267.
— P. SCHAEFFER, and E. L. WOLLMAN, 1960: Episomic elements in bacteria. Sympos. Soc. Gen. Microbiol., Cambridge: University Press **10**, 67.
— et E. L. WOLLMAN, 1958: Les épisomes, éléments génétiques ajoutés. C. R. Acad. Sci. (Paris) **247**, 154.
— — 1961: Sexuality and the genetics of bacteria. New York: Academic Press.
JACOBSON, W., and M. WEBB, 1952: The two nucleoproteins during mitosis. Exp. Cell Res. **3**, 163.
JAKOB, HEDWIG, 1950: Influence de la colchicine sur le développement de certaines algues d'eau douce. C. R. Acad. Sci. (Paris) **230**, 1203.
JOHNSON, C., 1912: The morphology and reactions of *Bacillus megatherium*. Cbl. Bakt. II **35**, 209.
JOHNSON, F. H., and D. H. GRAY, 1949: Nuclei and large bodies of luminous bacteria in relation to salt concentration, osmotic pressure, temperature and urethane. J. Bact. **58**, 675.
JOHNSON, T. B., and E. B. BROWN, 1922: The preparation of nucleic acid from the nucleoprotein of tubercle bacilli. The pyrimidines contained in tuberculinic acid. J. biol. Chem. **54**, 721, 731.
JORDAN, D. C., and I. GRINYER, 1965: Electron microscopy of the bacteroids and root nodules of *Lupinus luteus*. Canad. J. Microbiol. **11**, 721.

Jordan, D. O., 1951: Physicochemical properties of the nucleic acids. Progr. Biophys. 2, 51.

Joset, F., B. Low, and R. Krisch, 1964: Induction by radiation of a new direction of chromosome transfer during conjugation in an Hfr strain of *Escherichia coli*. Biochem. biophys. Res. Commun. 17, 742.

Jyssum, K., 1965: Polarity of chromosome replication in *Neisseria meningitidis*. J. Bact. 90, 1182.

— and S. Lie, 1965: Genetic factors as determining competence in transformation of *Neisseria meningitidis* I. A permanent loss of competence. Acta path. microbiol. scand. 63, 306.

Kadoya, M., H. Mitsui, Y. Takagi, E. Otaka, H. Suzuki, and S. Osawa, 1964: O deoxyribonucleic acid-protein complex having DNA-polymerase and RNA-polymerase activities in cell-free extracts of *Escherichia coli*. Biochim. biophys. Acta 91, 36.

Kaiser, A. D., 1965: Cohesion and the biological activity of bacteriophage lambda DNA. J. molec. Biol. 13, 78.

Kanazir, D., and M. Errera, 1959: Alterations of intracellular deoxyribonucleic acid and their biological consequence. Cold Spr. Harb. Symp. quant. Biol. 21, 19.

Kanemasa, Y., 1962: Selective staining of cytoplasmic membrane and nuclear apparatus of bacteria. Acta med. Okayama 16, 33.

Katz, S., 1963: The reversible reaction of Hg(II) and double-stranded polynucleotides. A step-function theory and its significance. Biochim. biophys. Acta 68, 240.

Kawata, T., 1958: Electron microscopy of ultrathin sections of *Micrococcus lysodeikticus*. Yonago Acta med. 3, 32.

–– 1961: Electron microscopy of the fine structure of *Corynebacterium diphtheriae*, with special reference to the intracytoplasmic membrane system. Jap. J. Microbiol. 5, 441.

— T. Inoue, and A. Takagi, 1963: Electron microscopy of spore formation and germination in *Bacillus subtilis*. Jap. J. Microbiol. 7, 23.

Kaye, A. M., R. Salomon, and B. Fridlender, 1967: Base composition and presence of methylated bases in DNA from a blue-green alga, *Plectonema boryanum*. J. molec. Biol. 24, 479.

Kellenberger, E., 1951/52: Die Einflüsse verschiedener Präparationsmethoden auf E. coli B. Z. wiss. Mikrosk. 60, 408.

— 1952 a: Les transformations des nucléoides de E. coli provoquées par les rayons ultra-violets. Experientia 8, 263.

–– 1952 b: Les nucléoides de *Escherichia coli* étudiés à l'aide du microscope électronique. Experientia 8, 99.

–– 1960: The physical state of the bacterial nucleus, in: Contributions to Microbial Genetics, 10th Sympos. Soc. Gen. Microbiol., Cambridge: University Press, 39.

–– 1961: Vegetative bacteriophage and the maturation of the virus particle. Adv. Virus Res. 8, 1.

— 1962: The study of natural and artificial DNA plasms by thin sections, in: The Interpretation of Ultrastructure. First Sympos. Int. Soc. Cell Biol. New York: Academic Press, 233.

— 1963: Organization of the genetic material in phage, bacteria and dinoflagellates, in: Genetics Today, Symposia held at the 11th Int. Congr. Genetics, s'Gravenhage, Oxford: Pergamon Press, vol. 2 of Proceedings, 309.

— K. G. Lark, and A. Bolle, 1962: Amino acid dependent control of DNA synthesis in bacteria and vegetative phage. Proc. nat. Acad. Sci. U.S.A. 48, 1860.

— et Antoinette Ryter, 1955: Contribution à l'étude du noyau bactérien. Schweiz. Z. allg. Path. 18, 1122.

— — 1956: Fixation et inclusion du matériel nucléaire de *Escherichia coli*. Experientia 12, 420.

–– — and Janine Séchaud, 1958 a: Electron microscope study of DNA-containing plasms II. Vegetative and mature phage DNA as compared with normal bacterial nucleoids in different physiological states. J. biophys. biochem. Cytol. 4, 671.

–– Janine Séchaud und Antoinette Ryter, 1958 b: Das Nucleoplasma der Bacteriennucleoide verglichen mit der DNS von vegetativen und reifen Phagen. 4. Int. Congr. Elektronenmikroskopie, Berlin: Springer 2, 212.

Kellenberger, Grete, et E. Kellenberger, 1956: Étude de souches colicinogènes au microscope électronique. Schweiz. Z. allg. Path. 19, 582.

KELLY, M. S., and R. H. PRITCHARD, 1965: Unstable linkage between genetic markers in transformation. J. Bact. **89**, 1314.

KIDSON, C., 1966: Deoxyribonucleic acid secondary structure in the region of the replication point. J. molec. Biol. **17**, 1.

KIM, KI-HAN, 1966: Properties and distribution of intracellular putrescine in a *Pseudomonas*. J. Bact. **91**, 193.

KINOSHITA, S., and S. ITAGAKI, 1959: Studies on the nucleus of the spore in *Streptomyces* I, II. Bot. Mag. **71**, 335.

KJELDGAARD, N. O., 1961: The kinetics of ribonucleic acid- and protein formation in *Salmonella typhimurium* during the transition between different states of balanced growth. Biochim. biophys. Acta **49**, 64.

— O. MAALØE, and M. SCHAECHTER, 1958: The transition between different physiological states during balanced growth of *Salmonella typhimurium*. J. gen. Microbiol. **19**, 607.

KLEINSCHMIDT, A., M. GEHATIA, und R. K. ZAHN, 1960 a: Über die molekulare Morphologie von Desoxyribonucleinsäuren. Kolloid-Z. **169**, 156.

— D. LANG, und R. K. ZAHN, 1960 b: Darstellung molekularer Fäden von Desoxyribonucleinsäuren. Naturwissenschaften **47**, 16.

— — — 1961: Über die intrazelluläre Formation von Bakterien-DNS. Z. Naturf. **16 b**, 730.

KLIENEBERGER-NOBEL, EMMY, 1945: Changes in the nuclear structure of bacteria, particularly during spore formation. J. Hyg. **44**, 99.

— 1947 a: The life-cycle of sporing *Actinomyces* as revealed by a study of their structure and septation. J. gen. Microbiol. **1**, 22.

— 1947 b: A cytological study of *Myxococcus*. J. gen. Microbiol. **1**, 33.

— 1947 c: Morphological appearances of various stages in *B. proteus* and *coli*. J. Hyg. **45**, 410.

— 1951: Filterable forms of bacteria. Bact. Rev. **15**, 77.

KNAYSI, G., 1942: The demonstration of a nucleus in the cell of a staphylococcus. J. Bact. **43**, 365.

— 1955 a: The structure, composition, and behavior of the nucleus of *Bacillus cereus*. J. Bact. **69**, 117.

— 1955 b: On the structure and nature of the endospore in strain C 3 of *Bacillus cereus*. J. Bact. **69**, 130.

— and R. F. BAKER, 1947: Demonstration, with the electron microscope, of a nucleus in *Bacillus mycoides* grown in a nitrogen-free medium. J. Bact. **53**, 539.

— J. HILLIER, and C. FABRICANT, 1951: The cytology of an avian strain of *Mycobacterium tuberculosis* studied with the electron and light microscope. J. Bact. **60**, 423.

KNÖLL, H., 1944: Zeiss-Nachr. **5**, 38 (cited after KNÖLL and ZAPF, 1952).

— and K. ZAPF, 1952: Untersuchungen zum Problem des Bakterienzellkerns, I. Mitteilung. Zbl. Bakt., I. Abt. Orig. **157**, 389.

— — 1954: Untersuchungen zum Problem des Bakterienzellkerns, II. Mitteilung. Zbl. Bakt., I. Abt. Orig. **161**, 241.

KNÖSEL, D., 1963: Karyologische Untersuchungen an sternbildenden Bakterien vermittels Färbe- und enzymatischer Verfahren, phasen- und elektronenoptisch. Zbl. Bakt., II. Abt. **116**, 113.

KOHIYAMA, M., H. LAMFROM, S. BRENNER et F. JACOB, 1963: Modifications de fonctions indispensables chez les mutants thermosensibles d'*Escherichia coli*. Sur une mutation empêchant la réplication du chromosome bactérien. C. R. Acad. Sci. (Paris) **257**, 1979.

KOHL, F. G., 1903: Über die Organisation und Physiologie der Cyanophyceenzelle und die mitotische Teilung ihres Kerns. Jena: Fischer.

KRAN, K., 1962: Cytologische Untersuchungen zur Entwicklung der Phagen T 2 und T 7 von *Escherichia coli*. Arch. Mikrobiol. **44**, 152.

— und F. W. SCHLOTE, 1959: Zur Darstellung des chromosomalen Materials in Bakterien. Arch. Mikrobiol. **34**, 412.

— — und H. G. SCHLEGEL, 1963: Cytologische Untersuchungen an *Chromatium okenii* Perty. Naturwissenschaften **50**, 728.

KRIEG, A., 1953: Fluoreszenzmikroskopischer Nachweis von Kernäquivalenten in Bakterien. Naturwissenschaften **40**, 414.

— 1954 a: Zur Zytologie und Enzymatik der Bakterien. Z. Naturf. **B 9**, 342.

— 1954 b: Existiert ein Nuclear-System in Bakterien? Zbl. Bakt., I. Abt. Orig. **161**, 212.

Krieg, A., 1954 c: Nachweis kernäuqivalenter Strukturen bei Bakterien *in vivo*, I—IV. Z. Hyg. 138, 357, 530; 139, 61, 64.
— 1954 d: Mikroskopische Untersuchungen *in vivo* an *Azotobacter*-Zellen. Naturwissenschaften 41, 147.
— 1954 e: Nachweis von Kernäquivalenten in Zyanophyzeen. Experientia 10, 204.
Krisch, R. E., 1965: Effects of radioactive phosphorus decay on genetic recombination in *E. coli*. K 12. Ph.D. Thesis, Pennsylvania U., Univ. Microfilm No. 65-1374.
— and M. J. Kvetkas, 1966: Inhibition of bacterial mating by amino acid deprivation. Biochem. biophys. Res. Commun. 22, 707.
Kruis, K., 1913: Mikrophotographie der Strukturen lebender Organismen, insbesondere der Bakterienkerne mit ultraviolettem Licht. Bull. int. de l'Acad. Sci. de Bohême.
Kubitschek, H. E., 1966: Mutation without segregation in bacteria with reduced dark repair ability. Proc. nat. Acad. Sci. U.S.A. 55, 269.
— H. E. Bendigkeit, and M. R. Loken, 1967: Onset of DNA synthesis in chemostat cultures of bacteria. Bact. Proc., 57.
Kuempel, P. L., M. Masters, and A. B. Pardee, 1965: Bursts of enzyme synthesis in the bacterial duplication cycle. Biochem. biophys. Res. Commun. 18, 858.
— and A. B. Pardee, 1963: The cycle of bacterial duplication. J. cell. comp. Physiol. 62, pt. II, suppl. 1, 15.
Kühlwein, H., und W. Rossner, 1963: Der „Zellkern" bei Myxobakterien. Naturwissenschaften 50, 339.
Kumar, H. D., 1962: Apparent genetic recombination in a blue-green alga, *Anacystis nidulans*. Nature 196, 1121.
— 1964: Effects of radiations on blue-green algae II. Effects on growth. Ann. Bot. (London) 28, 555.
— 1965: Effects of certain toxic chemicals and mutagens on the growth of the blue-green alga *Anacystis nidulans*. Canad. J. Bot. 43, 1523.
— 1966: Some experiments on the induction of nutritionally-deficient mutants in the blue-green alga *Anacystis nidulans*. Advancing Frontiers of Plant Sci. (India) 14, 133.
— H. N. Singh, and G. Prakash, 1967: The effect of proflavine on different strains of the blue-green alga *Anacystis nidulans*. Plant and Cell Physiol. 8, 171.
Kurland, C. G., M. Nomura, and J. D. Watson, 1962: The physical properties of the chloromycetin particles. J. molec. Biol. 4, 388.
Kushnarev, V. M., 1959: in Russian (Study of modifications of the bacterial cell during the preparation of ultra-thin sections). Mikrobiologija (Moscow) 28, 819.

Lacks, S., 1962: Molecular fate of DNA in genetic transformation in pneumococcus. J. molec. Biol. 5, 119.
Landman, O. E., and H. S. Ginoza, 1961 a: Genetic nature of stable L forms of *Salmonella paratyphi*. J. Bact. 81, 875.
— — 1961 b: Cytoplasmic and nuclear control of cell division and cell wall formation in bacteria. Genetics 46, 877.
Lang, D., und A. K. Kleinschmidt, 1964: Elektronenmikroskopische Darstellung, Längenverteilung und Konfiguration einsträngiger Desoxyribonucleinsäure im Cytochrom-c-Film. Biophysik 2, 73.
— — und R. K. Zahn, 1964: Konfiguration und Längenverteilung von DNA-Molekülen in Lösung. Biochim. biophys. Acta 88, 142.
Lark, Cynthia, 1966: Regulation of DNA synthesis in *Escherichia coli:* dependence of growth rates. Biochim. biophys. Acta 119, 517.
— and K. G. Lark, 1962: Cyclic deoxyribonucleic acid synthesis induced in *Escherichia coli* following the addition of ribonucleotides to the growth medium. Biochim. biophys. Acta 55, 401.
— — 1964: Evidence of two distinct aspects of the mechanism regulating chromosome replication in *Escherichia coli*. J. molec. Biol. 10, 120.
Lark, K. G., 1960: Studies on the mechanism regulating periodic DNA synthesis in synchronized cultures of *Alkaligenes fecalis*. Biochim. biophys. Acta 45, 121.
— 1961: Variation in bacterial acid-soluble deoxyribonucleotides during discontinuous deoxyribonucleic acid synthesis. Biochim. biophys. Acta 51, 107.
— 1963: Cellular control of DNA biosynthesis, in: Molecular Genetics, J. H. Taylor ed., New York: Academic Press, 1, 153.
— 1966: Regulation of chromosome replication and segregation in bacteria. Bact. Rev. 30, 3.

LARK, K. G., and R. E. BIRD, 1965 a: Segregation of the conserved units of DNA in *Escherichia coli*. Proc. nat. Acad. Sci. U.S.A. **54**, 1444.
— — 1965 b: Premature chromosome replication induced by thymine starvation: restriction of replication to one of the two partially completed replicas. J. molec. Biol. **13**, 607.
— and CYNTHIA LARK, 1965: Regulation of chromosome replication in *Escherichia coli*: alternate replication of two chromosomes at slow growth rate. J. molec. Biol. **13**, 105.
— O. MAALØE, and O. ROSTOCK. 1955: Cytological studies of nuclear division in *Salmonella typhimurium*. J. gen. Microbiol. **13**, 318.
— T. REPKO, and E. J. HOFFMAN, 1963: The effect of amino acid deprivation on subsequent deoxyribonucleic acid replication. Biochim. biophys. Acta **76**, 9.
LAZAROFF, N., and W. VISHNIAC, 1962: The participation of filament anastomosis in the developmental cycle of *Nostoc muscorum*, a blue-green alga. J. gen. Microbiol. **28**, 203.
LEA, D. E., 1955: Action of radiations on living cells, 2nd ed., Cambridge: University Press.
LEAK, L. V., 1964: Ultrastructure of a blue-green alga, *Anabaena* sp., and the incorporation of tritiated thymidine in the nucleoplasmic areas. J. Cell Biol. **23**, No. 2, 53 A.
— 1965: Electron microscopic autoradiography incorporation of ^{3}H-thymidine in a blue-green alga, *Anabaena* sp. J. Ultrastruct. Res. **12**, 135.
— and G. B. WILSON, 1965: Electron microscopic observations on a blue-green alga, *Anabaena* sp. Canad. J. Genet. Cytol. **7**, 237.
LEDERBERG, J., 1949: Aberrant heterozygotes in *Escherichia coli*. Proc. nat. Acad. Sci. U.S.A. **35**, 178.
— 1952: Cell genetics and hereditary symbiosis. Physiol. Rev. **32**, 403.
— 1956: Bacterial protoplasts induced by penicillin. Proc. nat. Acad. Sci. U.S.A. **42**, 574.
— 1957/58: Bacterial reproduction. Harvey Lectures 69.
— 1958: Extranuclear transmission of the F compatibility factor in *E. coli*. 7th Int. Congr. Microbiol. Stockholm, Absts., 59.
— L. L. CAVALLI, and E. M. LEDERBERG, 1952: Sex compatibility in *E. coli*. Genetics **37**, 720.
— and J. ST. CLAIR, 1958: Protoplast and L-type growth in *E. coli*. J. Bact. **75**, 143.
LEE, SYBIL, 1927: Cytological study of *Stigonema mammillosum*. Bot. Gaz. **83**, 420.
— 1960: Fixation of *E. coli* spheroplasts for electron microscopy. Exp. Cell Res. **21**, 252.
LESSLER, M. A., 1953: The nature and specificity of the Feulgen nucleal reaction. Int. Rev. Cytol. **2**, 231.
LEVINTHAL, C., and H. R. CRANE, 1956: On the unwinding of DNA. Proc. nat. Acad. Sci. U.S.A. **42**, 436.
— D. P. FAN, A. HIGA, and R. A. ZIMMERMANN, 1963: The decay and protection of messenger RNA in bacteria. Cold Spr. Harb. Symp. quant. Biol. **28**, 183.
LEWIS, I. M., 1941: The cytology of bacteria. Bact. Rev. **5**, 181.
LIEBERMEISTER, K., 1960: Morphology of the PPLO and L forms of *Proteus*. Ann. N.Y. Acad. Sci. **79**, 326.
LIQUORI, A. M., L. COSTANTINO, V. CRESCENZI, V. ELIA, E. GIGLIO, R. PULITT, M. DE SANTIS SAVINO, and V. VITAGLIANO, 1967: Complexes between DNA and polyamines: a molecular model. J. molec. Biol. **24**, 113.
LOUTIT, J. S., and L. E. PEARCE, 1965: Kinetics of mating of FP+ and FP-- strains of *Pseudomonas aeruginosa*. J. Bact. **90**, 425.
LURIA, S. E., and M. L. HUMAN, 1950: Chromatin staining of bacteria during bacteriophage infection. J. Bact. **59**, 551.
LUZZATTI, DENISE, 1966: Effect of thymine starvation on messenger ribonucleic acid synthesis in *Escherichia coli*. J. Bact. **92**, 1435.

MAALØE, O., 1961: The control of normal DNA replication in bacteria. Cold Spr. Harb. Symp. quant. Biol. **26**, 45.
— 1963: Role of protein synthesis in the DNA replication cycle in bacteria. J. cell. comp. Physiol. **62**, pt. II, suppl. 1, 31.
— 1964: *In vivo* replication of DNA. 6th Int. Congr. Biochem., New York, Abstracts I (Nucleic Acids), 17.

MAALØE, O., and A. BIRCH-ANDERSEN, 1956: On the organization of the "nuclear material" in *Salmonella typhimurium*, in "Bacterial Anatomy", 6th Symp. Soc. Gen. Microbiol., Cambridge: University Press, 261.

— — and F. S. SJÖSTRAND, 1954: Electron micrographs of sections of *E. coli* cells infected with the bacteriophage T 4. Biochim. biophys. Acta **15**, 12.

— and P. C. HANAWALT, 1961: Thymine deficiency and the normal DNA replication cycle. J. molec. Biol. **3**, 144.

McCARTHY, B. J., 1965: The evaluation of base sequences in polynucleotides. Progr. nucl. Acid Res. mol. Biol. **4**, 129.

— and E. T. BOLTON, 1964: Interaction of complementary RNA and DNA. J. molec. Biol. **8**, 184.

McCARTY, M., and O. T. AVERY, 1946: Studies on the chemical nature of the substance, inducing transformation in pneumococcal types II. Effect of desoxyribonuclease on the biological activity of the transforming substance III. An improved method for the isolation of the transforming substance and its application to *Pneumococcus*. J. exp. Med. **83**, 89, 97.

McFALL, ELIZABETH, and G. S. STENT, 1959: Continuous synthesis of deoxyribonucleic acid in *Escherichia coli*. Biochim. biophys. Acta **34**, 580.

McGEE, Z. A., M. ROGUL, S. FALKOW, and R. G. WITTLER, 1965: The relationship of *Mycoplasma pneumoniae* (Eaton agent) to *Streptococcus* MG: Application of genetic tests to determine relatedness of L forms and PPLO to bacteria. Proc. nat. Acad. Sci. U.S.A. **54**, 457.

McGREGOR, J. F., 1954: Nuclear division and the life cycle in a *Streptomyces* sp. J. gen. Microbiol. **11**, 52.

MAHLER, H. R., 1963: The interaction of nucleic acids with diamines. Biochim. biophys. Acta **68**, 211.

— B. D. MEHROTRA, and C. W. SHARP, 1960: Effects of diamines on the thermal transition of DNA. Biochem. biophys. Res. Commun. **4**, 79.

MAIER, S., 1964: A cytological study of *Thioploca ingrica* Wislouch. Ph.D. Thesis, Ohio State Univ.

— and R. G. E. MURRAY, 1964: The fine structure of *Thioploca ingrica* and a comparison with *Beggiatoa*. Canad. J. Microbiol. **11**, 645.

MALATJAN, M. N., 1963: in Russian (Modification of bacterial nuclei and mitochondria during development. Studies with the luminescence microscope). Mikrobiologija (Moscow) **32**, 806.

MALMGREN, B., and C.-G. HEDÉN, 1947: Studies of the nucleotide metabolism of bacteria III. The nucleotide metabolism of the Gram-negative bacteria. Acta path. microbiol. scand. **24**, 448.

MALMON, A. G., 1967: High resolution isotope tracing using induced nuclear reactions. Int. Conf. Use radioact. Isotopes in Pharmacology, Genève. Bruxelles: Presses Acad. Européennes.

MANDEL, M., and D. B. ROWLEY, 1963: Configuration and base composition of deoxyribonucleic acid from spores of *Bacillus subtilis* var. niger. J. Bact. **85**, 1445.

MARCOVICH, H., 1961: Recherches sur le mécanisme de l'action des rayons X chez *Escherichia coli* K 12 au moyen de la récombination génétique. Ann. Inst. Pasteur **101**, 660.

MARMUR, J., 1960: Thermal denaturation of deoxyribonucleic acid isolated from a thermophile. Biochim. biophys. Acta **38**, 342.

— 1963: New approaches to bacterial taxonomy. Ann. Rev. Microbiol. **17**, 329.

— and P. DOTY, 1961: Thermal renaturation of deoxyribonucleic acids. J. molec. Biol. **3**, 585.

— and C. M. GREENSPAN, 1963: Transcription *in vivo* of DNA from bacteriophage. Science **142**, 387.

— — E. PALECEK, F. M. KAHAN, J. LEVINE, and M. MANDEL, 1963: Specificity of the complementary RNA formed by *Bacillus subtilis* infected with bacteriophage SP 8. Cold Spr. Harb. Symp. quant. Biol. **28**, 191.

— and D. LANE, 1960: Strand separation and specific recombination in deoxyribonucleic acids: biological studies. Proc. nat. Acad. Sci. U.S.A. **46**, 453.

— R. ROWND, S. FALKOW, L. S. BARON, C. SCHILDKRAUT, and P. DOTY, 1961 a: The nature of intergeneric episomal infection. Proc. nat. Acad. Sci. U.S.A. **47**, 972.

— C. L. SCHILDKRAUT, and P. DOTY, 1961 b: The reversible denaturation of DNA and its use in studies of nucleic acid homologies and the biological relatedness of microorganisms. J. Chim. phys. Phys.-chim. biol. **58**, 943.

MARSHAK, A., 1951 a: Chromosome structure in *Escherichia coli*. Exp. Cell Res. 2, 243.
— 1951 b: Structures in *E. coli* resembling chromonemata. Proc. nat. Acad. Sci. U.S.A. 37, 38.
MARTIN, D., et F. JACOB, 1962: Transfer de l'épisome sexuel d'*Escherichia coli* à *Pasteurella pestis*. C. R. Acad. Sci. (Paris) 254, 3589.
MARUYAMA, Y., 1963: Biochemical aspects of the cell growth of *Escherichia coli* as studied by the method of synchronous culture. J. Bact. 72, 821.
— and K. G. LARK, 1961: Periodic synthesis of bacterial nucleic acids in the absence of protein synthesis. Exp. Cell Res. 25, 161.
MARVIN, D. A., and H. HOFFMANN-BERLING, 1963: Physical and chemical properties of two new small bacteriophages. Nature 197, 517.
MARX, RUTH, und W. HEUMANN, 1963: Elektronenmikroskopische Untersuchungen an Feinstrukturen der Bakterienzelle. Zeiss-Mitteilungen 3, 110.
MASON, D. J., and D. M. POWELSON, 1956: Nuclear division as observed in live bacteria by a new technique. J. Bact. 71, 474.
MASSIE, H. R., and B. H. ZIMM, 1965: Molecular weight of the DNA in the chromosomes of *E. coli* and *B. subtilis*. Proc. nat. Acad. Sci. U.S.A. 54, 1636.
MASTERS, M., and W. D. DONACHIE, 1966: Repression and the control of cyclic enzyme synthesis in *Bacillus subtilis*. Nature 209, 476.
— P. L. KUEMPEL, and A. B. PARDEE, 1964: Enzyme synthesis in synchronous cultures of bacteria. Biochem. biophys. Res. Commun. 15, 38.
— and A. B. PARDEE, 1965: Sequence of enzyme synthesis and gene replication during the cell cycle of *Bacillus subtilis*. Proc. nat. Acad. Sci. U.S.A. 54, 64.
MASUI, M., T. IWATA, A. ISHIMITSU, and Y. UMEBAYASHI, 1962: Desoxyribonucleoprotein of halophilic *Achromobacter*. Biochim. biophys. Acta 55, 384.
MATNEY, T. S., E. P. GOLDSCHMIDT, N. S. ERWIN, and R. A. SCROGGS, 1964: A preliminary map of genomic sites for F-attachment in *Escherichia coli* K 12. Biochem. biophys. Res. Commun. 17, 278.
MAXWELL, R. E., and V. S. NICKEL, 1954: Filament formation in *Escherichia coli* induced by azaserine and other antineoplastic agents. Science 120, 270.
MEHROTRA, B. D., 1964: Some studies on DNA-small ion interactions. Ph.D. Thesis Indiana U., Univ. Microfilms No. 64-12060.
MESELSON, M., and F. W. STAHL, 1958 a: The replication of DNA in *Escherichia coli*. Proc. nat. Acad. Sci. U.S.A. 44, 671.
— — 1958 b: The replication of DNA. Cold Spr. Harb. Symp. quant. Biol. 23, 9.
MESSER, W., 1963: Zur Struktur und Vermehrungsweise des Chromosoms von *Escherichia coli*. Z. Naturf. 18 b, 385.
METZ, A., 1958: Beiträge zur Morphologie und Biologie der Lithium- und Cäsiumformen der *Salmonella paratyphi B*. Zbl. Bakt., I. Abt. Orig. 171, 461.
MEYER, A., 1904: Orientierende Untersuchungen über Verbreitung, Morphologie und Chemie des Volutins. Bot. Ztg. 62, 113.
— 1912: Die Zelle der Bakterien. Jena: G. Fischer.
MEYER, H., 1965: Die Feinstruktur des Kernäquivalentes von Bakterien bei der „Negativfärbung" mit Phosphowolframsäure. Naturwissenschaften 52, 437.
MINCK, R., J.-P. EBEL et A. MINCK, 1950: La choix du fixateur en cytologie bactérienne. C. R. Soc. Biol. 144, 288.
— et A. MINCK, 1948: Le noyau des bactéries anaérobies. C. R. Soc. Biol. 142, 531.
MOCKERIDGE, F. A., 1926: An examination of *Nostoc* for nuclear materials. Brit. J. exp. Biol. 4, 301.
MOORE, R. B., 1966: Survival studies on the blue-green alga, *Anacystis nidulans*. Ph.D. Thesis, Univ. Texas, Austin.
MORGAN, C., H. S. ROSENKRANZ, and H. S. CARR, 1967: Effect of chloramphenicol on *Escherichia coli* as observed by the electron microscope. Bact. Proc., 111.
MOSES, M. J., 1956: Chromosomal structures in crayfish spermatocytes. J. biophys. biochem. Cytol. 2, 215.
MÓTUZOVA, I. A., 1963: in Russian (On the fine structure of the L form of *Proteus vulgaris*). Mikrobiologija (Moscow) 32, 61.
MUDD, S., and A. G. SMITH, 1950: Electron and light microscopic studies of bacterial nuclei. I. Adaptation of cytological processing to electron microscopy; bacterial nuclei as vesicular structures. J. Bact. 59, 561.
MUNCH-PETERSEN, A., and J. NEUHARD, 1964: Studies on the acid-soluble nucleotide pool in thymine-requiring mutants of *Escherichia coli* during thymine starvation I. Accumulation of deoxyadenine triphosphate in *Escherichia coli* 15 T—A—U—. Biochim. biophys. Acta 80, 542.

Murray, R. G. E., 1953: The problem of fixation for studies of bacterial nuclei. 6th Int. Congr. Microbiol. Roma, Sympos. Citologia Batterica (Roma: Istituto Superiore di Sanità), 136.
— and H. C. Douglas, 1950: The reproductive mechanism of *Rhodomicrobium vannielii* and the accompanying nuclear changes. J. Bact. **59**, 157.

Nadson, G., 1895: Über den Bau des Cyanophyceenprotoplasten. Scripta bot. St. Petersburg **4**.
Nagata, T., 1962: Polarity and synchrony in the replication of DNA molecules of bacteria. Biochem. biophys. Res. Commun. **8**, 348.
— 1963 a: The molecular synchrony and sequential replication of DNA in *Escherichia coli*. Proc. nat. Acad. Sci. U.S.A. **49**, 551.
— 1963 b: The sequential replication of *E. coli* DNA. Cold Spr. Harb. Symp. quant. Biol. **28**, 55.
Nakada, D., 1960: Involvement of newly-formed protein in the synthesis of deoxy-ribonucleic acid. Biochim. biophys. Acta **44**, 241.
— and F. J. Ryan, 1961: Replication of deoxyribonucleic acid in non-dividing bacteria. Nature **189**, 398.
— E. Strelzoff, R. Rudner, and F. J. Ryan, 1960: Is DNA replication a necessary condition for mutation? Z. Vererbungsl. **91**, 210.
Nermut, M. V., 1959: Staining of nuclear substance in microorganisms after alkaline hydrolysis. Scripta med. (Brno) **32**, 45.
— 1962: Cytological observations on large bodies and protoplasts of bacteria. II. Nuclear equivalents and their fusion. Folia microbiol. (Praha) **7**, 47.
Neugnot, Denise, 1950: Contribution à l'étude cytochimique des Cyanophycées par application des techniques de mise en évidence de l'appareil nucléaire chez les bactéries. C. R. Acad. Sci. (Paris) **230**, 1311.
Neuhard, J., and A. Munch-Petersen, 1966: Studies on the acid-soluble nucleotide pool in thymine-requiring mutants of *Escherichia coli* during thymine starvation. II. Changes in the amounts of deoxycytidine triphosphate and deoxyadenosine triphosphate in *Escherichia coli* 15 T⁻–A⁻–U⁻. Biochim. biophys. Acta **114**, 61.
Neumann, F., 1930: Die Frage nach dem Kern der Bakterien und ihre Beantwortung mit Hilfe der Nuklealfärbung. Berl. Münch. tierärzt. Wschr. **46**, 101.
— 1941: Untersuchungen zur Erforschung der Kernverhältnisse bei den Bakterien. Zbl. Bakt., II. Abt. **103**, 385.
Nisman, B., et H. Fukuhara, 1960: Rôle de l'acide deoxyribonucléique dans la synthèse de protéine par les deux fractions enzymatiques liées à des particules de *Escherichia coli*. C. R. Acad. Sci. (Paris) **250**, 410.
Nomura, M., and K. Hosokawa, 1965: Biosynthesis of ribosomes. Fate of chloramphenicol particles and of pulse-labeled RNA in *Escherichia coli*. J. molec. Biol. **12**, 242.

Ogg, J. E., and R. D. Humphrey, 1963: Small-cell segregants from a possibly homozygous diploid strain of *Escherichia coli*. J. Bact. **85**, 801.
— and M. R. Zelle, 1957: Isolation and characterization of a large cell possibly diploid strain of *E. coli*. J. Bact. **74**, 477.
Oishi, M., H. Yoshikawa, and N. Sueoka, 1964: Synchronous and dichotomous replication of the *Bacillus subtilis* chromosome during spore germination. Nature **204**, 1069.
Okazaki, R., T. Okazaki, and Y. Kuriki, 1959: Incorporation of [³H]thymidine in a deoxyriboside-requiring bacterium. Biochim. biophys. Acta **33**, 289.
Okazaki, T., and R. Okazaki, 1959: Studies of deoxyribonucleic acid synthesis and cell growth in the deoxyriboside requiring bacteria *Lactobacillus acidophilus*. II. Deoxyribonucleic acid synthesis in relation to ribonucleic acid synthesis and protein synthesis. Biochim. biophys. Acta **35**, 434.
Olive, E. W., 1905: Mitotic division of the nuclei of the Cyanophyceae. Beih. Bot. Cbl. **18**, 9.
Onofrio, F. D., e G. Falcone, 1957: Sull'influenza di un agente chelante sulla divisione batterica. Riv. Ist. sieroter. ital. **32**, 279.
Ørskov, I., and F. Ørskov, 1960: An antigen termed F⁺ occurring in F⁺ *E. coli* strains. Acta path. microbiol. scand. **48**, 37.
Ottolenghi, E., and R. D. Hotchkiss, 1960: Appearance of genetic transforming activity in pneumococcal cultures. Science **132**, 1257.

PACHLER, P. F., A. L. KOCH, and M. SCHAECHTER, 1965: Continuity of DNA synthesis in *Escherichia coli*. J. molec. Biol. **11**, 650.

PAINTER, R. B., F. FORRO Jr., and W. L. HUGHES. 1958: Distribution of tritium-labelled thymidine in *Escherichia coli* during cell multiplication. Nature **181**, 328.

— and H. S. GINOZA, 1966: Some characteristics of the resistance transfer factor (RTF) episome as determined by inactivation with tritium, P^{32} and gamma radiation. Biophys. J. **6**, 153.

PAKULA, R., and W. WALCZAK, 1963: On the nature of competence of transformation in streptococci. J. gen. Microbiol. **31**, 125.

PALADE, G. E., 1952: A study of fixation for electron microscopy. J. exp. Med. **95**, 285.

PALMADE, CHRISTIANE, 1961: Acide désoxyribonucléique et des désoxyribonucléo-protéines bactériennes. Thèse Fac. Sci. Univ. Strasbourg.

— and COLETTE VENDRELY, 1956: Sur une méthode nouvelle d'extraction de l'acide désoxyribonucléique des bactéries. C. R. Acad. Sci. (Paris) **242**, 2870.

PANKRATZ, H. S., and C. C. BOWEN, 1963: Cytology of blue-green algae. I. The cells of *Symploca muscorum*. Amer. J. Bot. **50**, 387.

PAPPAS, G. D., 1956: Helical structures in the nucleus of *Amoeba proteus*. J. biophys. biochem. Cytol. **2**, 221.

PARDEE, A. B., and L. S. PRESTIDGE, 1956: The dependence of nucleic acid synthesis on the presence of amino acids in *Escherichia coli*. J. Bact. **71**, 677.

PARK, I. W., and J. DE LEY, 1967: Ancestral remnants in deoxyribonucleic acid from *Pseudomonas* and *Xanthomonas*. Antonie v. Lecuwenhoek **33**, 1.

PARVIS, D., 1950: Comportamento del nucleo batterico nella lisi di batteriofago. Boll. Soc. ital. Biol. sper. **26**, 1.

— 1954: Sulle modificazioni provocate nei nuclei batterici dai cosiddetti „veleni mitotici". Boll. Ist. sieroter. milan. **33**, 463.

PAULING, C., and P. HANAWALT, 1965: Non-conservative DNA replication in bacteria after thymine starvation. Proc. nat. Acad. Sci. **54**, 1728.

PAYNE, J. I., P. E. HARTMAN, S. MUDD, and A. W. PHILLIPS, 1956: Cytological analysis of ultraviolet-irradiated *Escherichia coli*. III. Reactions of a sensitive strain and its resistant mutants. J. Bact. **72**, 461.

PERRET, C. J., 1958: The effect of growth rate on the anatomy of *Escherichia coli*. J. gen. Microbiol. **18**. No. 1, vii.

PESHKOFF, M. A., 1946: Fine structure and mechanism of division of the nuclei of the bacterium *Caryophanon latum*. Nature **157**, 137.

— and I. A. MOTUZOVA, 1963: in Russian (Fine structure of M and L forms of bacteria). Mikrobiologija (Moscow) **32**, 799.

PETERS, D., and R. WIGAND, 1953: Enzymatisch-elektronenoptische Analyse der Nucleinsäureverteilung, dargestellt an *Escherichia coli* als Modell. Z. Naturf. B **8**, 180.

PETIT, A., 1927: Contribution à l'étude cytologique et taxonomique des bactéries. Thèse, Fac. Sci. Paris.

PETTER, Mlle H. F. M., 1933 a: Structure d'une Sarcine: *Sarcina gigantea*. C. R. Acad. Sci. (Paris) **196**, 300.

— 1933 b: La réaction nucléale de Feulgen chez quelques végétaux inférieurs. C. R. Acad. Sci. (Paris) **197**, 88.

PETTIJOHN, D. E., and P. C. HANAWALT, 1964: Isolation of partially density-labelled DNA molecules from bacteria. J. molec. Biol. **8**, 170.

PHILLIPS, O. P., 1904: A comparative study of the cytology and movements of the Cyanophyceae. Contrib. bot. Lab. Univ. Pennsylvania **2**, 237.

PIÉCHAUD, M., 1949: Coloration des corps chromatiques des Entérobactériacées par un colorant neutre sans hydrolyse préalable. Ann. Inst. Pasteur **76**, 66.

— 1951: Coloration directe du noyau des bactéries et des levures. Bull. Micr. appl. Sér. 2, **1**, 178.

— 1954: La coloration sans hydrolyse du noyau des bactéries. Ann. Inst. Pasteur **86**, 787.

PIEKARSKI, G., 1937: Cytologische Untersuchungen an Paratyphus- und Colibakterien. Arch. Mikrobiol. **8**, 428.

— 1938: Zytologische Untersuchungen an Bakterien im ultravioletten Licht. Zbl. Bakt., I. Abt. Orig. **142**, 70.

— 1939: Lichtoptische und übermikroskopische Untersuchungen zum Problem des Bakterienzellkerns. Zbl. Bakt., I. Abt. Orig. **144**, 140.

— 1940: Über kernähnliche Strukturen bei *Bacillus mycoides* Flügge. Arch. Mikrobiol. **11**, 406.

Piekarski, G., 1950: Haben Bakterien einen Zellkern? (Zur Definition des Zellkerns.) Naturwissenschaften 37, 201.
— und P. Giesbrecht, 1955: Diskussionsbemerkung zu: E. Kellenberger et Antoinette Ryter, Contribution à l'étude du noyau bactérien. Schweiz. Z. allg. Path. 18, 1136.
— — 1956: Zytologische Untersuchungen an Bacillus megaterium mit Hilfe ultradünner Schnitte. Naturwissenschaften 43, 89.
— — und P. Janssen, 1967: Über Probleme bei der Anwendung von Acridinorange als Methode zur fluoreszenzmikroskopischen Anfärbung von Toxoplasmen. Arch. Hyg. Bakt. 151, 550.
— und G. Pontieri, 1956: Über Feinstrukturen in den kernähnlichen Körpern von E. coli. Zbl. Bakt., I. Abt. Orig. 165, 242.
Pitzurra, M., e G. Ragni, 1958: Aspetti dismorfici nelle microculture di E. coli provenienti da culture vecchie. Boll. Ist. sieroter. milan. 37, 262.
Pochon, J., Y. T. Tchan, et T. L. Wang, 1948: Recherches sur le cycle morphologique et l'appareil nucléaire des Azotobacter. Ann. Inst. Pasteur 74, 182.
Poindexter, Jeanne S., 1964: Biological properties and classification of the Caulobacter group. Bact. Rev. 28, 231.
— and Germaine Cohen-Bazire, 1964: The fine structure of stalked bacteria belonging to the family Caulobacteriaceae. J. Cell Biol. 23, 587.
Poljansky, G., und G. Petruschewsky, 1929: Zur Frage über die Struktur der Cyanophyceenzelle. Arch. Protistenk. 67, 11.
Pontefract, R. D., and F. S. Thatcher, 1965: A cytological study of normal and radiation-resistant Escherichia coli. Canad. J. Microbiol. 11, 271.
Pontieri, G., 1956: Über die sog. „large bodies" von E. coli. Zbl. Bakt., I. Abt. Orig. 165, 524.
Powell, E. O., 1956: Growth rate and generation time, with special reference to continuous culture. J. gen. Microbiol. 15, 492.
Preuner, R., 1951: Untersuchungen über den Bakterienkern. Arch. Hyg. Bakt. 134, 281.
— 1953: Untersuchungen zur Feinstruktur ruhender Bacillensporen. Zbl. Bakt., I. Abt. Orig. 160, 227.
— und Jutta von Prittwitz und Gaffron, 1951: Über feulgenpositive Körper in Bazillen und Bakterien und ihr Verhalten gegenüber Streptomycin und Penicillin. Zbl. Bakt., I. Abt. Orig. 157, 244.
Preusser, H.-J., 1958: Elektronenmikroskopische Untersuchungen über die Cytologie von Proteus vulgaris. Arch. Mikrobiol. 29, 17.
— 1959: Form und Größe des Kernäquivalentes von Escherichia coli in Abhängigkeit von den Kulturbedingungen. Arch. Mikrobiol. 33, 105.
Prévot, A. R., et C. Mazurek, 1953: Recherches sur la substance nucléaire de Bifidibacterium bifidum et d'Actinobacterium cellulitis. Ann. Inst. Pasteur 85, 125.
— et M. Reymond, 1948: Étude cytologique des sphéroides de Spherophorus funduliformis. Ann. Inst. Pasteur 74, 334.
Pringsheim, E. G., 1949: The relationship between bacteria and Myxophyceae. Bact. Rev. 13, 47.
— 1953: Die Stellung der grünen Bakterien im System der Organismen. Arch. Mikrobiol. 19, 353.
— 1960: Der grüne Farbstoff der Chlorobakterien, eine Berichtigung. Arch. Mikrobiol. 36, 98.
— and C. F. Robinow, 1946: Observations on two very large bacteria, Caryophanon latum Peshkoff and Lineola longa (nomen provisorium). J. gen. Microbiol. 1, 267.
Pritchard, R. H., and K. G. Lark, 1964: Induction of replication by thymine starvation at the chromosome origin in Escherichia coli. J. molec. Biol. 9, 288.
Ptashne, M., 1965: Replication and host modification of DNA transferred during bacterial mating. J. molec. Biol. 11, 829.
Pulvirenti, G. B., 1951: L'azione filamentizzante dell'urea sulla cellula batterica. Riv. Biol. N. S. 43, 319.

Quersin, L., 1950: Images nucléaires et coloration de Gram. Ann. Inst. Pasteur 79, 767.

Rajchert-Trzpil, M., 1965: The influence of chloramphenicol on the conjugation yield in E. coli K 12. Bull. Acad. polon. Sci., Ser. Sci. biol. 13, 211.
Rasmussen, R. E., and R. B. Painter, 1963: On the early onset of thymineless death occurring after exposure to ultraviolet light. Biochim. biophys. Acta 76, 157.

REICHENOW, E., 1928: Ergebnisse mit der Nuclealfärbung bei Bakterien. Arch. Protistenk. **61**, 142.

REMSEN, C. C., D. G. LUNDGREN, and R. A. SLEPECKY, 1966: Inhibition of development of the spore septum and membranes in *Bacillus cereus* by β-phenethyl alcohol. J. Bact. **91**, 324.

RHOADES, J. V., J. ABELSON, T. C. PINKERTON, and C. A. THOMAS Jr., 1965: The question of interruptions in bacteriophage DNA molecules. Biophys. Soc. 9th Ann. Meeting, 96.

RIS, H., 1956: A study of chromosomes with the electron microscope. J. biophys. biochem. Cytol. **2**, Suppl. 385.

— 1961: Ultrastructure and molecular organization of genetic systems. Canad. J. Genet. Cytol. **3**, 95.

— 1962: Interpretation of ultrastructure in the cell nucleus. 1st Sympos. Int. Soc. Cell Biol. (New York: Academic Press 1964), 69.

— and B. L. CHANDLER, 1963: The ultrastructure of genetic systems in *Prokaryotes* and *Eukaryotes*. Cold. Spr. Harb. Symp. quant. Biol. **28**, 1.

— and J. Fox, 1949: The cytology of Rickettsiae. J. exp. Med. **89**, 681.

— and R. N. SINGH, 1961: Electron microscope studies on blue-green algae. J. biophys. biochem. Cytol. **9**, 63.

RITCHIE, A. E., and I. N. ROBINSON, 1967: Electron microscopic anatomy of a rumen spirochaete. Bact. Proc., 25.

ROBERTS, R. B., 1960: Synthetic aspects of ribosomes. Ann. N.Y. Acad. Sci. **88**, 752.

ROBINOW, C. F.: 1942: A study of the nuclear apparatus of bacteria. Proc. Roy. Soc. (London) **B 130**, 299.

— 1944: Cytological observations on *Bact. coli, Proteus vulgaris* and various aerobic spore-forming bacteria with special reference to the nuclear structures. J. Hyg. **43**, 413.

— 1949 a: Nuclear apparatus and cell structure of rod-shaped bacteria, addendum to: R. J. DUBOS: The Bacterial Cell. Cambridge, Mass.: Harvard University Press.

— 1949 b: Cytological observations on bacteria. Exp. Cell Res. Suppl. **1**, 204.

— 1953 a: Spore structure as revealed by thin sections. J. Bact. **66**, 300.

— 1953 b: On the structure of bacterial spores. 6th Int. Congr. Microbiol., Symposium Citologia Batterica. Roma: Istituto Superiore di Sanità, 139.

— 1956: The chromatin bodies of bacteria. Bact. Rev. **20**, 207.

— 1957: Kurzer Hinweis auf *Metabacterium polyspora*. Z. Tropenmed. Parasit. **8**, 225.

— 1960: Outline of the visible organization of bacteria, in: The Cell. J. BRACHET and A. E. MIRSKY ed., New York: Academic Press, **4**, pt. 1, 45.

— 1962: Morphology of the bacterial nucleus. Brit. med. Bull. **18**, 31.

— and V. E. COSSLETT, 1948: Nuclei and other structures of bacteria. J. appl. Physiol. **19**, 124.

— and C. L. HANNAY, 1953: The nuclear structures of *B. megaterium*. 6th Int. Congr. Microbiol. Roma, Riassunti delle Communicazione **1**, 67.

ROGOLSKY, M., and R. A. SLEPECKY, 1964: Elimination of a genetic determinant for sporulation of *Bacillus subtilis* with acriflavin. Biochem. biophys. Res. Commun. **16**, 204.

ROGUL, M., Z. A. McGEE, R. G. WITTLER, and S. FALKOW, 1965: Nucleic acid homologies in selected bacteria, L forms and *Mycoplasma species*. J. Bact. **90**, 1200.

ROLFE, R., 1962: The molecular arrangement of the conserved subunits of DNA. J. molec. Biol. **4**, 22.

— 1963: Changes in the physical state of DNA during the replication cycle. Proc. nat. Acad. Sci. U.S.A. **49**, 386.

— 1967: On the mechanism of thymineless death in *Bacillus subtilis*. Proc. nat. Acad. Sci. U.S.A. **57**, 114.

ROLLY, H., 1951: Der Entwicklungszyklus eines aeroben Sporenbildners (*Bac. sphaericus*) unter besonderer Berücksichtigung der Nuclealstrukturen. Zbl. Bakt., I. Abt. Orig. **157**, 407.

ROPARS, C., et R. VIOVY, 1964: Fixation de l'ion cuivrique sur l'acide désoxyribonucléique de thymus de veau. Influences de la force ionique et du pH. C. R. Acad. Sci. (Paris) **258**, 731.

ROSENBERG, B. H., and L. F. CAVALIERI, 1964: On the transient template for *in vivo* DNA synthesis. Proc. nat. Acad. Sci. U.S.A. **51**, 826.

ROSENBERG, E., M. KATARSKI, and P. GOTTLIEB, 1967: Deoxyribonucleic acid synthesis during exponential growth and microcyst formation in *Myxococcus*. J. Bact. **93**, 1402.

Ross, P. D., and R. L. Scruggs, 1964: Electrophoresis of DNA. II. Specific inter-action of univalent and divalent cations with DNA. Biopolymers 2, 79.

Rossner, W., 1963: Der Einfluß von Streptomycin auf Cyanophyceen. II. Elektronen-mikroskopische Untersuchungen an *Phormidium minnesotense* (Tilden) Drouet. Planta 60, 166.

Roth, T. F., and D. R. Helinski, 1967: Circular forms of a bacterial plasmid. Bact. Proc. 49.

Rouillier, C., et E. Fauré-Fremiet, 1957: Les constituants cytoplasmiques d'une bactérie sulfureuse *Theovolum maius* Hinze. C. R. Soc. Biol. 151, 903.

Rudner, R., Mrs. B. Prokop-Schneider, and E. Chargaff, 1964: Rhythmic alternations in the rate of synthesis and the composition of rapidly labelled ribonucleic acid during the synchronous growth of bacteria. Nature 203, 479.

— E. Rejman, and E. Chargaff, 1965: Genetic implications of periodic pulsations of the rate of synthesis and the composition of rapidly labeled bacterial RNA. Proc. nat. Acad. Sci. U.S.A. 54, 904.

Růzička, V., 1913: Eine Methode zur Darstellung der Struktur fertiger Bakterien-sporen, nebst Bemerkung über das Reifen derselben. Cbl. Bakt. II 36, 577.

Ryan, F. J., und K. Kiritani, 1959: Effect of temperature on natural mutation in *Escherichia coli.* J. gen. Microbiol. 20. 644.

Ryter, Antoinette, 1958: Étude morphologique de la sporulation de *Bacillus sub-tilis.* Ann. Inst. Pasteur 108, 40.

— 1960: Étude au microscope électronique des transformations nucléaires de *E. coli* K^{12}S and K^{12}S (λ 26) après irradiation aux rayons ultraviolets et aux rayons X. J. biophys. biochem. Cytol. 8, 399.

— 1967: Relationship between synthesis of the cytoplasmic membrane and nuclear segregation in *Bacillus subtilis.* Fol. microbiol. (Praha) 12, 283.

— 1968: Association of the nucleus and the membrane of bacteria: a morphological study. Bact. Rev. 32, 39.

— B. Bloom, et J.-P. Aubert, 1966 a: Localisation intracellulaire des acides ribo-nucléiques synthétisés pendant la sporulation chez *Bacillus subtilis.* C. R. Acad. Sci. (Paris) D 262, 1305.

— et F. Jacob, 1963: Étude au microscope électronique des relations entre méso-somes et noyaux chez *Bac. subtilis.* C. R. Acad. Sci. (Paris) 257, 3060.

— — 1964: Étude au microscope électronique de la liaison entre noyau et mésosome chez *Bacillus subtilis.* Ann. Inst. Pasteur 107, 384.

— — 1966 a: Étude morphologique de la liaison du noyau à la membrane chez *E. coli* et chez les protoplastes de *B. subtilis.* Ann. Inst. Pasteur 110, 801.

— — 1966 b: Ségrégation des noyaux chez *Bacillus subtilis* au cours de la germi-nation des spores. C. R. Acad. Sci. (Paris) D 263, 1176.

— — 1967: Ségrégation des noyaux pendant la croissance et la germination de *B. subtilis.* C. R. Acad. Sci. (Paris) D 264, 2254.

— and E. Kellenberger, 1957: Fixation of the "nucleus" of *E. coli* bacteria. 15th Ann. Meeting Electron Microsc. Soc. Amer., M.I.T. Cambridge, Mass., Abstracts, 26.

— — 1958: Étude au microscope électronique de plasmas contenant de l'acide désoxyribonucléique. I. Les nucléoides des bactéries en croissance active. Z. Naturf. 13 b, 597.

— and O. E. Landman, 1964: Electron microscope study of the relationship between mesosome loss and the stable L state (or protoplast state) in *Bacillus subtilis.* J. Bact. 88, 457.

— et M. Piéchaud, 1963: Étude au microscope électronique de quelques souches de *Moraxella.* Ann. Inst. Pasteur 105, 1071.

— P. Schaeffer, et H. Ionesco, 1966 b: Classification cytologique, par leur stade de blocage, des mutants de sporulation de *Bacillus subtilis* Marburg. Ann. Inst. Pasteur 110, 305.

Saito, H., and Y. Ikeda, 1957: Radiokinetic evidence for the bipartite or diploid nucleus in conidia of *Streptomyces griseoflavus.* J. gen. appl. Microbiol. 3, 250.

— — 1958: The life-cycle of *Streptomyces griseoflavus.* Cytologia 23, 496.

— — 1959: Cytogenetic studies on *Streptomyces griseoflavus.* Ann. N.Y. Acad. Sci. 81, 862.

Salvatore, G., e G. Pontieri, 1956: Osservazioni dul dimorfismo di *Proactinomyces ruber.* Giorn. Microbiol. 1, 385.

Sassuchin, D., 1935: Zum Studium der Protisten- und Bakterienkerne. Mitt. I: Über die Nuclealreaktion und ihre Anwendung bei protozoologischen und bakteriologischen Untersuchungen. Arch. Protistenk. 84, 186.

Schachtele, C. F., 1965: Canavanine death in Escherichia coli. J. molec. Biol. 14, 474.

Schaechter, M., 1961 a: Patterns of cellular control during unbalanced growth. Cold Spr. Harb. Symp. quant. Biol. 26, 53.

— 1961 b: Discussion to O. Maaløe: The control of normal DNA replication in bacteria. Cold Spr. Harb. Symp. quant. Biol. 26, 52.

— M. W. Bentzon, and O. Maaløe, 1959: Synthesis of deoxyribonucleic acid during the division cycle of bacteria. Nature 183, 1207.

— and V. O. Laing, 1961: Direct observation of fusion of bacterial nuclei. J. Bact. 81, 667.

— O. Maaløe, and N. O. Kjeldgaard, 1958: Dependency on medium and temperature of cell size and chemical composition during balanced growth of Salmonella typhimurium. J. gen. Microbiol. 19, 592.

Schaeffer, P., J. Millet, and J.-P. Aubert, 1965: Catabolic repression of bacterial sporulation. Proc. nat. Acad. Sci. U.S.A. 54, 704.

Schlossberger, H., A. Jakob und G. Piekarski, 1950: Zur systematischen Stellung der Spirochäten. Naturwissenschaften 37, 186.

Schmitz, F., 1879: Untersuchungen über die Zellkerne der Thallophyten. Sitzber. niederrhein. Ges. Nat. Heilk. (Bonn), 345.

Schreil, W., 1961: Vergleichende Elektronenmikroskopie reiner DNS und der DNS des Bakteriennukleoids. Experientia 17, 391.

— 1964 (Schreil, W. H.): Studies on the fixation of artificial and bacterial DNA plasms for the electron microscopy of thin sections. J. Cell Biol. 22, 1.

Schuler, R., 1952: Die vitale Darstellung der Chromatinstrukturen von Bact. coli mit Hilfe des Fluoreszenzmikroskops. Naturwissenschaften 39, 90.

Schweisfurth, R., 1959: Einwirkung einiger Gifte auf Zell- und Nucleoidteilung bei Bakterien. Zbl. Bakt., II. Abt. 113, 48.

Sedar, A. W., and R. W. Burde, 1965: The demonstration of the succinic dehydrogenase system in Bacillus subtilis using tetranitro-blue tetrazolium combined with techniques of electron microscopy. J. Cell Biol. 27, 53.

Sermonti, G., and I. Spada-Sermonti, 1955: Genetic recombination in Streptomyces. Nature 176, 121.

— — 1959: Genetics of Streptomyces coelicolor. Ann. N.Y. Acad. Sci. 81, 854.

Setlow, J. K., and R. B. Setlow, 1960: Evidence for the existence of a single-stranded stage of T 2 bacteriophage during replication. Proc. nat. Acad. Sci. U.S.A. 46, 791.

Setlow, R. (B.), 1960: The use of action spectra to determine the physical state of DNA in vivo. Biochim. biophys. Acta 39, 180.

— and R. Boyce, 1961: The ultraviolet light inactivation of Φ X 174 bacteriophage at different wave lengths and pH's. Biophys. J. 1, 29.

Shack, J., and B. S. Bynum, 1959: Determination of the interaction of deoxyribonucleate and magnesium ions by means of a metal ion indicator. Nature 184, 635.

Shadomy, S., 1963: Cytological studies of bacteriophage infection in the mycobacteria. Ph.D. Thesis, Univ. Calif. Los Angeles.

Shakulov, R. S., A. A. Bogdanov, and A. S. Spirin, 1963: in Russian (Reconstruction of ribosomal particles from chloromycetin particles and protein in vitro). Dokl. Akad. Nauk. S.S.S.R. 153, 223.

Shamina, Z. B., 1964: in Russian (Observations on nuclear elements during sporogenesis of actinomycetes). Mikrobiologija (Moscow) 33, 831.

Shinke, N., and K. Ueda, 1956: A cytomorphological and cytochemical study of Cyanophyta. I. An electron microscope study of Oscillatoria princeps. Mem. Coll. Sci. Univ. Kyoto, Ser. B 23, 101.

Silver, S. D., 1963: The transfer of material during mating in Escherichia coli. Transfer of DNA and upper limits in the transfer of RNA and protein. J. molec. Biol. 6, 349.

Singh, H. N., 1967: Genetic control of sporulation in the blue-green alga Anabaena doliolum Bharadwaja. Planta 75, 33.

Singh, R. N., and H. N. Singh, 1964: Ultra-violet induced mutants of blue-green algae. Arch. Mikrobiol. 48, 109, 118.

— and R. Sinha, 1965: Genetic recombination in a blue-green alga, Cylindrospermum maius Kuetz. Nature 207, 782.

Skoczylas, O., 1954: Über die Mitose in der Peridineengattung Ceratium. Dissertation, Fak. f. Chemie, TH Darmstadt.

Smith, A. G., 1950: Electron and light microscopic studies of bacterial nuclei. II. An improved staining technique for the nuclear chromatin of bacterial cells. J. Bact. **59**, 575.

Smith, D. W., and P. C. Hanawalt, 1965: State of aggregation of the growing point in the bacterial chromosome. Biophys. Soc. ninth annu. Meetg. 162.

Smith, M. A., M. Salas, W. M. Stanley, A. J. Wahba, and S. Ochoa, 1966: Direction of reading of the genetic message. II. Proc. nat. Acad. Sci. U.S.A. **55**, 141.

Smith, P. F., 1964: Comparative physiology of pleuropneumonia-like and L-type organisms. Bact. Rev. **28**, 97.

Smithies, W. R., and N. E. Gibbons, 1955: The deoxyribose nucleic acid slime layer of some halophilic bacteria. Canad. J. Microbiol. **1**, 614.

Sneath, P. H. A., and J. Lederberg, 1961: Inhibition by periodate of mating in *Escherichia coli*. Proc. nat. Acad. Sci. U.S.A. **47**, 86.

Soška, J., and K. G. Lark, 1966: Regulation of nucleic acid synthesis in *Lactobacillus acidophilus* R-26. Biochim. biophys. Acta **119**, 526.

Spearing, J. K., 1937: Cytological studies on the myxophyceae. Arch. Protistenk. **89**, 209.

— 1961: Studies on the cyanophycean cell. I. Vital staining. A study in the production of artifacts. La Cellule **61**, 241.

Spiegelman, S., 1959: Protein and nucleic acid synthesis in subcellular fractions of bacterial cells, in: Recent Progress in Microbiology, Symposia held at the 7th Int. Congr. Microbiol. Stockholm, 82.

— A. L. Aronson, and P. C. Fitz-James, 1958: Isolation and characterization of nuclear bodies from protoplasts of *Bacillus megaterium*. J. Bact. **75**, 102.

— B. D. Hall, and R. Storck, 1961: The occurrence of natural DNA-RNA complexes in *E. coli* infected with T 2. Proc. nat. Acad. Sci. U.S.A. **47**, 1135.

Spirin, A. S., A. N. Belozerskij, D. G. Kudlai, A. G. Skavronskaja, and V. G. Mitereva, 1958 a: in Russian (Composition of nucleic acids in the formation of saccharolytically inert forms of intestinal bacteria). Biokhimija **23**, 154.

— A. G. Skavronskaja, and A. Pretel-Martines, 1958 b: Nucleic acid content in ageing cultures of colon bacilli. Microbiologija (English Translation) **27**, 271.

Srinivasan, V. R., 1966: Sporogen, an " inductor" for bacterial cell differentiation. Nature **209**, 537.

Stacey, M., 1953: Chemistry of the Gram staining and of the Feulgen and Dische reactions for nuclear material. Nature **171** 507.

Stanier, R. Y., 1961: La place des bactéries dans le monde vivant. Ann. Inst. Pasteur **101**, 297.

— and C. B. van Niel, 1962: The concept of a bacterium. Arch. Mikrobiol. **42**, 17.

Stapp, C., 1942: Der Pflanzenkrebs und sein Erreger *Pseudomonas tumefaciens*. XI. Mitteilung: Zytologische Untersuchung des bakteriellen Erregers. Zbl. Bakt., II. Abt. **105**, 1.

— und D. Knösel, 1954: Zur Genetik sternbildender Bakterien. Zbl. Bakt., II. Abt. **108**, 243.

— — 1956 a: Phasenoptisch-cytologische Untersuchungen sternbildender Bakterien. Zbl. Bakt., II. Abt. **109**, 25.

— — 1956 b: Fortgeführte Untersuchungen über den Entwicklungszyklus und die Karyologie sternbildender Bakterien. Zbl. Bakt., II. Abt. **109**, 416.

— — 1956 c: Zur Frage des Vorkommens sogenannter „Polyzygoten" bei sternbildenden Bakterien. Zbl. Bakt., II. Abt. **109**, 473.

Stempen, H., 1950: Demonstration of the chromatine bodies of *E. coli* and *Proteus vulgaris* with the aid of the phase contrast microscope. J. Bact. **60**, 81.

Stent, G. S., and C. R. Fuerst, 1960: Genetic and physiological effects of the decay of incorporated radioactive phosphorus in bacterial viruses and bacteria. Advance biol. med. Phys. **7**, 1.

— and O. Maaløe, 1953: Radioactive phosphorus tracer studies on the reproduction of T 4 bacteriophage. II. Kinetics of phosphorus assimilation. Biochim. biophys. Acta **10**, 55.

Stern, K. G., and M. A. Steinberg, 1953: Desoxyribonucleic acid complexes of rare earths. Biochim. biophys. Acta **11**, 553.

Stille, B., 1937: Zytologische Untersuchungen an Bakterien mit Hilfe der Feulgenschen Nuclealreaktion. Arch. Mikrobiol. **8**, 124.

Stoker, M. G. P., K. M. Smith, and P. Fiset, 1956: Internal structure of *Rickettsia burnetii* as shown by electron microscopy of thin sections. J. gen. Microbiol. **15**, 632.

Stonehill, E. H., 1965: On the mechanism of DNA replication. J. theor. Biol. **9**, 323.

STOUGHTON, R. H., 1930: The morphology and cytology of *Bacterium malvacearum.* Proc. Roy. Soc. (London) B 105, 469.
— 1932: The morphology and cytology of *Bacterium malvacearum* E. F. S. Part II. Reproduction and cell-fusion. Proc. Roy. Soc. (London) B 111, 46.
STOUTHAMER, A. H., and P. G. DE HAAN, 1963: Kinetics of F-curing by acridine orange. J. gen. Microbiol. 31, xii.
STUY, J. H., 1958: The nucleic acids of *Bacillus cereus.* J. Bact. 76, 179.
SUEOKA, N., and H. YOSHIKAWA, 1963: Regulation of chromosome replication in *Bacillus subtilis.* Cold Spr. Harb. Symp. quant. Biol. 28, 47.
SUIT, J. C., 1963: Localization of deoxyribonucleic acid-like ribonucleic acid in a "membrane" fraction of *Escherichia coli.* Biochim. biophys. Acta 72, 488.
SURINGAR, W. F. R., 1865: De sarcine (Sarcina ventriculi Goodsir) Leeuwarden: G. T. N. Suringar.
— 1866: Ein Wort über den Zellenbau der Sarcina. Bot. Ztg. 24, 269, 277.
SUZUKI, H., J. PANGBORN, and W. W. KILGORE, 1967: Filamentous cells of *Escherichia coli* formed in the presence of mitomycin. J. Bact. 93, 683.
SZABO, G., T. VALYI-NAGY, and S. VITALES, 1963: An endogenic factor regulating the life cycle of *Streptomyces griseus.* Excerpta med., Int. Congr. Ser., No. 59, 88.
SZYBALSKI, W., and D. H. BRAENDLE, 1956: Genetic recombination in *Streptomyces.* Bact. Proc., 48.

TABOR, H., 1961: The stabilization of *Bacillus subtilis* transforming principle by spermine. Biochem. biophys. Res. Commun. 4, 228.
— C. W. TABOR, and S. M. ROSENTHAL, 1961: The biochemistry of the polyamines: spermidine and spermine. Annu. Rev. Biochem. 30, 579.
TAKAGI, A., T. KAWATA, S. YAMAMOTO, T. KUBO, and S. OKITA, 1960: Electron microscopic studies on ultrathin sections of spores of *Clostridium tetani* and *Clostridium histolyticum*, with special reference to sporulation and spore germination process. Jap. J. Microbiol. 4, 137.
TAKAHASHI, I., and N. E. GIBBONS, 1957: Effect of salt concentration on the extracellular nucleic acids of *Micrococcus halodenitrificans.* Canad. J. Microbiol. 3, 687.
TAMURA, A., 1959: Studies on the structure of *Rickettsia orientalis* as shown by chemical treatment. Kyoto Univ. Inst. Virus Res., Annu. Rept., Ser. B, 187.
TANAKA, K., 1961: Studies on the Cytology of *Spirillum,* I., II. Scient. Papers Coll. gen. Educ. Univ. Tokyo 11, 69, 83.
TAYLOR, A. L., and E. A. ADELBERG, 1960: Linkage analysis with very high frequency males of *Escherichia coli.* Genetics 45, 1233.
THOMAS, J.-A., 1932: Contribution à l'étude cytologique des Schizophytes. Arch. zool. exp. gén. 72, 417.
TICHONENKO, A. S., I. A. BESPELOVA, and A. S. KRIVISKI, 1964: Electron microscopy studies of RNA phage MS 2 and its reproduction in bacterial cells. Proc. 3rd Europ. Reg. Conf. Electron Microscopy Prague, vol. B, 545.
TOCCHINI-VALENTINI, G. P., M. STODOLSKY, A. AURISICCHIO, M. SARNAT, F. GRAZIOSI, S. B. WEISS, and E. P. GEIDUSCHEK, 1963: On the asymmetry of RNA synthesis *in vivo.* Proc. nat. Acad. Sci. U.S.A. 50, 935.
TOMASZ, A., 1965: Control of competent state in *Pneumococcus* by a hormone-like cell product: an example for a new type of regulatory mechanism in bacteria. Nature 208, 155.
— J. D. JAMIESON, and E. OTTOLENGHI, 1964: The fine structure of *Diplococcus pneumoniae.* J. Cell Biol. 22, 453.
TOMIZAWA, J., 1960: Genetic structure of recombinant chromosomes formed after mating in *Escherichia coli* K 12. Proc. nat. Acad. Sci. U.S.A. 46, 91.
TOMLIN, S. G., and J. W. MAY, 1955: Electron microscopy of sectioned bacteria. A study of *Escherichia coli.* Austral. J. exp. Biol. Med. 33, 249.
TOSCHI, G., 1963/64: Concentrazione intracellulare e composizione delle poli-ammine in microorganismi mesofili e termofili. Boll. Soc. ital. Biol. sper. 39, 2125.
TREICK, R. W., and W. A. KONETZKA, 1964: Physiological state of *Escherichia coli* and the inhibition of deoxyribonucleic acid synthesis by phenethyl alcohol. J. Bact. 88, 1580.
TRONNIER, E. A., 1952: Zur Existenz des sog. Karyoid-Systems bei *Corynebacterium diphtheriae.* Zbl. Bakt., I. Abt. Orig. 159, 213.
— 1953: Zur Darstellung der Nukleoide bei Bakterien mit Azur-Eosin. Naturwissenschaften 40, 511.
TSCHENZOFF, B., 1916: Die Kernteilung bei *Euglena viridis.* Arch. Protistenk. 36, 137.

Tsumita, T., and E. Chargaff, 1958: Studies on nucleoproteins. VI. The deoxyribonucleoprotein and the deoxyribonucleic acid of bovine tubercle bacilli (BCG).
Biochim. biophys. Acta 29, 568.
Tuffery, A. A., 1954a: Systematic position of the genus Oscillospira. Nature
174, 838.
— 1954b: The nuclear structures of Oscillospira guilliermondi Chatton and Perard.
J. gen. Microbiol. 10, 342.
— 1955: Nuclear changes in the growth cycle of Caryophanon latum. Exp. Cell
Res. 9, 182.
Tulasne, R., 1949a: Sur la cytologie des bactéries vivantes étudiées grace au
microscope à contraste de phase. C. R. Soc. Biol. 143, 1390.
— 1949b: Données nouvelles sur la cytologie des bactéries apportées par la microscopie de contraste de phase. C. R. Soc. Biol. 143, 1392.
— 1953: Le cycle L et les formes naines des bactéries. Sixth Int. Congr. Microbiol.
Roma, Symposium Citologia Batterica. Roma: Istituto Superiore di Sanità, 144.
— et R. Minck, 1947: Mise en évidence des noyaux des cocci par la ribonucléase.
C. R. Soc. Biol. 141, 1255.
— and R. Vendrely, 1947a: Demonstration of bacterial nuclei with ribonuclease.
Nature 160, 225.
— — 1947b: Mise en évidence des noyaux bactériens par la ribonucléase. C. R.
Soc. Biol. 141, 674.
— — et R. Minck, 1948a: Noyaux bactériens et pénicilline. C. R. Soc. Biol. 142, 237.
— — — 1948b: Noyaux bactériens et streptomycine. C. R. Soc. Biol. 142, 1012.

Vahlkampf, E., 1905: Beiträge zur Biologie und Entwicklungsgeschichte von Amoeba
limax einschließlich der Züchtung auf künstlichen Nährböden. Arch. Protistenk.
5, 67.
Valentine, R. C., 1966: Sexual differentiation of E. coli. Biochem. biophys. Res.
Commun. 22, 156.
Van Baalen, C., 1965: Mutation of the blue-green alga, Anacystis nidulans. Science
149, 70.
van Iterson, Woutera, 1960: Membranes, particular organelles and peripheral bodies
in bacteria. Proc. Europ. Regional Conf. Electron Microscopy, Delft, 2, 763.
— 1962: Membraneous structures in micro-organisms, in: Recent Progress in Microbiology, Symposia held at the Eight Int. Congr. Microbiol., Montréal. Toronto:
Univ. Toronto Press, 14.
— and C. F. Robinow, 1961: Observations with the electron microscope on the fine
structure of the nuclei of two spherical bacteria. J. biophys. biochem. Cytol.
9, 171.
van Tubergen, R. P., 1961: The use of radioautography and electron microscopy
for the localization of tritium label in bacteria. J. biophys. biochem. Cytol.
9, 219.
— and R. B. Setlow, 1961: Quantitative autoradiographic studies on exponentially
growing cultures of Escherichia coli. The distribution of parental DNA, RNA,
protein, and cell wall on progeny cells. Biophys. J. 1, 589.
Vejdovský, F., 1900: Bemerkungen über den Bau und die Entwicklung der Bakterien. Cbl. Bakt. II 6, 577.
— 1904: Über den Kern der Bakterien und seine Teilung. Cbl. Bakt. II 11, 481.
Vendrely, R., 1953: Histochimie du noyau et du cytoplasme bactérien. 6th Int.
Congr. Microbiol. Roma, Symposium Citologia Batterica. Roma: Istituto Superiore di Sanità, 82.
— et Jeanne Lipardy, 1946: Acides nucléiques et noyaux bactériens. C. R. Acad.
Sci. (Paris) 223, 342.
Venner, H., and C. Zimmer, 1964: Effekte von Schwermetallionen auf den Helix-
Knäuel-Übergang der Desoxyribonucleinsäure. Naturwissenschaften 51, 173.
— — 1966: Studies on nucleic acids. VIII. Changes in the stability of DNA
secondary structure by interaction with divalent metal ions. Biopolymers 4, 321.
Vielmetter, W., und W. Messer, 1964: Zum Wachstum des Bakterienchromosoms
(E. coli K 12, Hfr und F−). Ber. Bunsenges. phys. Chem. 68, 742.
Vinter, V., and R. A. Slepecky, 1965: Direct transition of outgrowing bacterial
spores to new sporangia without intermediate cell division. J. Bact. 90, 803.
Voelz, H., 1966: The fate of the cell envelope of Myxococcus xanthus during microcyst germination. Arch. Mikrobiol. 55, 110.
— and M. Dworkin, 1962: Fine structure of Myxococcus xanthus during morphogenesis. J. Bact. 84, 943.

Voelz, H., Ursula Voelz, and R. O. Ortigoza, 1966: The "polphosphate overplus" phenomenon in *Myxococcus xanthus* and its influence on the architecture of the cell. Arch. Mikrobiol. 53, 371.

Voit, K., 1925: Über das Verhalten der Bakterien zur Nuclealfärbung. Z. ges. exp. Med. 47, 183.

— 1927: Über das Verhalten der Bakterien zur Nuclealfärbung. II. Mitteilung. Z. ges. exp. Med. 55, 564.

von Plotho, O., 1942: Die chromatische Substanz bei Aktinomyceten. Arch. Mikrobiol. 11, 285.

von Zastrow, Eva-Maria, 1953: Über die Organisation der Cyanophyceenzelle. Arch. Mikrobiol. 19, 174.

Wager, H., 1901: Cytology of the Cyanophyceae. Rep. Brit. Ass. 830.

Wake, R. G., 1963: Sequential replication of DNA in synchronously germinating *Bacillus subtilis* spores. Biochem. biophys. Res. Commun. 13, 67.

Walker, J. R., and A. B. Pardee, 1967: Relation between DNA and septum formation in *Escherichia coli*. Bact. Proc., 56.

Watanabe, T., 1963: Infective heredity of multiple drug resistance in bacteria. Bact. Rev. 27, 87.

Watson, J. D., and F. H. C. Crick, 1953 a: A structure for deoxyribose nucleic acid. Nature 171, 737.

— — 1953 b: Genetical implications of the structure of deoxyribonucleic acid. Nature 171, 964.

Webb, R. B., J. B. Clark, and H. I. Chance, 1954: A cytological study of *Nocardia corallina* and other actinomycetes. J. Bact. 67, 498.

Weed, L. L., 1963: Effects of copper on *Bacillus subtilis*. J. Bact. 85, 1003.

Weibull, C., 1965: Structure of bacterial L forms and their parent bacteria. J. Bact. 90, 1467.

Welsch, M., et E. Nihoul, 1948: A propos de la mise en évidence du noyau bactérien. C. R. Soc. Biol. 142, 1449.

Weygand, F., A. Wacker, und H. Dellweg, 1951: Spaltung von Desoxyribonuclein-säure mit Bleihydroxyd und Isolierung der Desoxyribose durch kontinuierliche Gegenstromverteilung. Z. Naturf. B 6, 130.

Whitfield, J. F., 1962: Lysogeny. Brit. med. Bull. 18, 56.

— and R. G. E. Murray, 1954: A cytological study of the lysogenization of *Shigella dysenteriae* with P 1 and P 2 bacteriophages. Canad. J. Microbiol. 1, 216.

— — 1956: The effects of the ionic environment on the chromatin structures of bacteria. Canad. J. Microbiol. 2, 245.

— — 1957: Observations on the initial cytological effects of bacteriophage infection. Canad. J. Microbiol. 3, 493.

Wiberg, J. S., and W. F. Neumann, 1957: The binding of bivalent metals by deoxy-ribonucleic and ribonucleic acids. Arch. Biochem. Biophys. 72, 66.

Wigand, R., and D. Peters, 1954 a: Licht- und elektronenmikroskopische Untersuchungen über den Abbau gramnegativer Kokken mit Nucleasen und Proteasen. Z. Naturf. B 9, 586.

— — 1954 b: Abbauversuche an *Haemobartonella muris* und *Eperythrozoon coccoides*. Z. Tropenmed. Parasit. 5, 482.

Wilkins, M. H. F., and G. Zubay, 1959 a: X-ray diffraction studies of the molecular structure of nucleohistone and chromosomes. J. molec. Biol. 1, 179.

— — 1959 b: The absence of histone in the bacterium *Escherichia coli*. II. X-ray diffraction of nucleoprotein extract. J. biophys. biochem. Cytol. 5, 55.

Wilkinson, J. F., and J. P. Duguid, 1960: The influence of cultural conditions on bacterial cytology. Int. Rev. Cytol. 9, 1.

Williams, Marion A., 1959: Chromatin patterns in *Spirillum anulus*. J. gen. Microbiol. 21, 109.

— and S. C. Rittenberg, 1957: Nuclear fusion in *Spirillum* spp. J. gen. Microbiol. 17, 274.

Winkler, Annelise, und Mechthild Knoch, 1951: Zur Fixation von Bakterienpräparaten. Zbl. Bakt., I. Abt. Orig. 157, 239.

— — und H. König, 1951: Vergleichende licht- und elektronenmikroskopische Untersuchungen an Bakterien. Naturwissenschaften 38, 241.

Witkin, Evelyn M., 1951: Nuclear segregation and the delayed appearance of induced mutants in *Escherichia coli*. Cold Spr. Harb. Symp. quant. Biol. 16, 357.

— 1958: Post-irradiation metabolism and the timing of ultraviolet-induced mutations in bacteria. Proc. 10th Int. Congr. Genetics Montreal, 280.

WOESE, C. R., 1958: Comparison of the X-ray sensitivity of bacterial spores. J. Bact. **75**, 5.

WOHLFARTH-BOTTERMANN, K.-E., 1961: Cytologische Studien VII, Strukturaspekte der Grundsubstanz des Cytoplasmas nach Einwirkung verschiedener Fixierungsmittel. Untersuchungen am Hyaloplasma von Amöben der Limax- und Proteus-Gruppe. Protoplasma **53**, 259.

WOLFE, S. L., M. BEER, and C. R. ZOBEL, 1962: The relative staining of nucleic acids in a model system and in tissue. Proc. Fifth Int. Congr. Electron microscopy, Philadelphia. New York: Academic Press, **2**, 0—6.

WOLLMAN, E. L., et F. JACOB, 1955: Sur le mécanisme du transfert de matériel génétique au cours de la recombinaison chez *Escherichia coli* K 12. C. R. Acad. Sci. (Paris) **240**, 2449.

— — 1959: La sexualité des bactéries. Paris: Masson & Cie.

— — and W. HAYES, 1956: Conjugation and genetic recombination in *Escherichia coli* K 12. Cold Spr. Harb. Symp. quant. Biol. **21**, 141.

WON, W. D., 1950: The production of "giant" cells of *Pasteurella pestis* by treatment with camphor. J. Bact. **60**, 102.

WYATT, G. R., and S. S. COHEN, 1952: Nucleic acids of Rickettsiac. Nature **170**, 846.

WYSS, O., M. G. NEUMANN, and M. D. SOCOLOFSKY, 1961: Development and germination of the *Azotobacter* cyst. J. biophys. biochem. Cytol. **10**, 555.

YAMANE, T., and N. DAVIDSON, 1961: On the complexing of desoxyribonucleic acid (DNA) by mercuric ion. J. amer. chem. Soc. **83**, 2599.

— — 1962: On the complexing of deoxyribonucleic acid by silver (I). Biochim. biophys. Acta **55**, 609.

YOSHIKAWA, H., 1965: DNA synthesis during germination of *Bacillus subtilis* spores. Proc. nat. Acad. Sci. U.S.A. **53**, 1476.

— A. O'SULLIVAN, and N. SUEOKA, 1964: Sequential replication of the *Bacillus subtilis* chromosome. III. Regulation of initiation. Proc. nat. Acad. Sci. U.S.A. **52**, 973.

— and N. SUEOKA, 1963: Sequential replication of the *Bacillus subtilis* chromosome. I. Comparison of marker frequencies in exponential and stationary growth phases. Proc. nat. Acad. Sci. U.S.A. **49**, 559.

YOUNG, I. ELIZABETH, and P. C. FITZ-JAMES, 1959: Pattern of synthesis of deoxyribonucleic acid in *Bacillus cereus* growing synchronously out of spores. Nature **183**, 372.

YUASA, A., T. HIRANO, and O. SUZUKI, 1955: in Japanese (Studies in the cytology of *Mycobacterium*. III. The digestion-method of the nucleus). Jap. J. Genet. **30**, 78.

ZAPF, K., 1957: Untersuchungen zum Problem des Bakterienzellkerns Penicillinbeeinflußter Coli-Bakterien. Zbl. Bakt., I. Abt. Orig. **167**, 565.

ZAVARZIN, G. A., 1960: in Russian (Developmental cycle and nuclear apparatus of *Hyphomicrobium vulgare* Stutz. et Hartleb). Mikrobiologija (Moscow) **29**, 38.

— 1961: in Russian (The budding bacteria). Mikrobiologija (Moscow) **30**, 952.

ZETTNOW, E., 1899: Romanowski's Färbung bei Bakterien. Z. Hyg. **30**, 1.

ZOBEL, C. R., and M. BEER, 1961: Electron stains. I. Chemical studies on the interaction of DNA with uranyl salts. J. biophys. biochem. Cytol. **10**, 335.

ZUBAY, G., 1959: The interaction of nucleic acid with Mg-ions. Biochim. biophys. Acta **32**, 233.

— and P. DOTY, 1958: Nucleic acid interactions with metal ions and amino acids. Biochim. biophys. Acta **29**, 47.

— and M. R. WATSON, 1959: The absence of histone in the bacterium *Escherichia coli*. I. Preparation and analysis of nucleoprotein extract. J. biophys. biochem. Cytol. **5**, 51.

Author Index

Page numbers in *italics* refer to illustrations